Inverse Gas Chromatography

ACS SYMPOSIUM SERIES **391**

Inverse Gas Chromatography

Characterization of Polymers and Other Materials

Douglas R. Lloyd, EDITOR
The University of Texas at Austin

Thomas Carl Ward, EDITOR
Virginia Polytechnic Institute and State University

Henry P. Schreiber, EDITOR
Ecole Polytechnique de Montréal

Clara C. Pizaña, ASSOCIATE EDITOR

Developed from a symposium sponsored
by the Division of Polymeric Materials: Science and Engineering
of the American Chemical Society
and the Macromolecular Science & Engineering Division
of the Chemical Institute of Canada
at the Third Chemical Congress of North America
(195th National Meeting of the American Chemical Society),
Toronto, Ontario, Canada,
June 5–11, 1988

 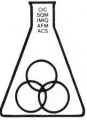

American Chemical Society, Washington, DC 1989

CHEM
sap/ae

Library of Congress Cataloging-in-Publication Data

Inverse gas chromatography.

(ACS Symposium Series, ISSN 0097–6156; 391).

"Developed from a symposium sponsored by the
Division of Polymeric Materials: Science and Engineering
of the American Chemical Society and the Macromolecular
Science & Engineering Division of the Chemical Institute
of Canada at the Third Chemical Congress of North
America (195th National Meeting of the American
Chemical Society), Toronto, Ontario, Canada, June 5–11,
1988."

Papers presented at the Symposium on Polymer
Characterization by Inverse Gas Chromatography.

Includes bibliographies and indexes.

1. Inverse gas chromatography—Congresses.
2. Polymers—Analysis—Congresses.

I. Lloyd, Douglas R., 1948–   . II. Ward, Thomas C.,
1941–   . III. Schreiber, Henry P., 1926–   .
IV. American Chemical Society. Division of Polymeric
Materials: Science and Engineering. V. Chemical Institute
of Canada. Macromolecular Science Division.
VI. Chemical Congress of North America (3rd: 1988:
Toronto, Ont.) VII. American Chemical Society. Meeting
(195th: 1988: Toronto, Ont.) VIII. Symposium on
Polymer Characterization by Inverse Gas Chromatography
(1988: Toronto, Ont.) IX. Series.

QD79.C45I59   1989   543'.0896   89–6628

   ISBN 0–8412–1610–X

sd
8/18/89
JJ

Foreword

The ACS SYMPOSIUM SERIES was founded in 1974 to provide a medium for publishing symposia quickly in book form. The format of the Series parallels that of the continuing ADVANCES IN CHEMISTRY SERIES except that, in order to save time, the papers are not typeset but are reproduced as they are submitted by the authors in camera-ready form. Papers are reviewed under the supervision of the Editors with the assistance of the Series Advisory Board and are selected to maintain the integrity of the symposia; however, verbatim reproductions of previously published papers are not accepted. Both reviews and reports of research are acceptable, because symposia may embrace both types of presentation.

Contents

SURFACE AND INTERFACE CHARACTERIZATION

ANALYTICAL APPLICATIONS

SPECIAL APPLICATIONS

INDEXES

Preface

INVERSE GAS CHROMATOGRAPHY (IGC) is a useful technique for characterizing synthetic and biological polymers, copolymers, polymer blends, glass and carbon fibers, coal, and solid foods.

The technique involves creating within a column a stationary phase of the solid material of interest. The stationary phase may be a thin polymeric coating on an inert substrate, a finely divided solid, or a thin polymeric coating on the column wall. A volatile *probe* of known characteristics is passed through the column via an inert mobile phase and the output is monitored. The residence time of the probe and the shape of the chromatogram indicate the characteristics of the stationary phase and its interaction with the probe. Thus, IGC is a variation of conventional gas chromatography.

IGC can be used to determine various properties of the stationary phase, such as the transition temperatures, polymer–polymer interaction parameters, acid–base characteristics, solubility parameters, crystallinity, surface tension, and surface area. IGC can also be used to determine properties of the vapor–solid system, such as adsorption properties, heat of adsorption, interaction parameters, interfacial energy, and diffusion coefficients. The advantages of IGC are simplicity and speed of data collection, accuracy and precision of the data, relatively low capital investment, and dependability and low operating cost of the equipment.

Increased interest in IGC has resulted in a dramatic increase in the number of papers on the subject. In the decade following the first mention of IGC in 1967, approximately 30 papers were published about IGC. In the ensuing decade, more than 300 IGC papers were published. This book, the first to focus exclusively on IGC, contains 19 of the 20 papers presented at the Symposium on Polymer Characterization by IGC. Three chapters were added to broaden the scope of this volume.

Following an overview of this volume, the first section, which consists of three chapters, focuses on methodology and instrumentation. The next three sections consider characterization of vapor–polymer systems (4 chapters), polymer–polymer systems (4 chapters), and surfaces and interfaces (6 chapters). The final two sections cover analytical applications (2 chapters) and the application of IGC in coal characterization and food science (1 chapter each).

Each chapter of the volume was critiqued by at least two scientists (in addition to the editors) and revised accordingly by the authors. The editors appreciate the assistance provided by the reviewers and by Cheryl Shanks of the ACS Books Department. Finally, the editors gratefully acknowledge the magnificent job done by our associate editor, Clara C. Pizaña, in copy editing manuscripts and assisting in the final production stages of the book.

DOUGLAS R. LLOYD
The University of Texas
Austin, TX 78712–1062

THOMAS CARL WARD
Virginia Polytechnic Institute and State University
Blacksburg, VA 24061

HENRY P. SCHREIBER
Ecole Polytechnique
Montréal, Québec H3C 3A7, Canada

December 7, 1988

Chapter 1

Overview of Inverse Gas Chromatography

Henry P. Schreiber[1] and Douglas R. Lloyd[2]

[1]Department of Chemical Engineering, Ecole Polytechnique, P.O. Box 6079, Station A, Montréal, Québec H3C 3A7, Canada
[2]Department of Chemical Engineering, Center for Polymer Research, University of Texas, Austin, TX 78712

Inverse gas chromatography (IGC) is an extension of conventional gas chromatography (GC) in which a non-volatile material to be investigated is immobilized within a GC column. This *stationary phase* is then characterized by monitoring the passage of volatile *probe* molecules of known properties as they are carried through the column via an inert gas. The invention of IGC in 1967 (1) and the subsequent development of IGC theory and methodology, beginning in 1976 (2) and continuing today, are the consequence of the increasing interest in materials science. While IGC was initially used only in the study of synthetic polymers, today, as evidenced in this book, IGC is used to study synthetic and biological polymers, copolymers, polymer blends, glass and carbon fibers, coal, and solid foods.

Laub (3) estimates that in the decade prior to 1977, IGC related publications contributed only approximately 3% of the total of some 1400 devoted to the overall subject of gas chromatography. In an update, soon to be published (3), Laub notes that in the decade prior to 1987, the total number of GC papers has remained about the same, but IGC now accounts for some 300 of these, or nearly 30% of the total. These statistics alone suggest the desirability of symposia in which the most recent advances in theoretical and practical aspects of the IGC methodology are represented.

Reasons for IGC's higher profile in the technical literature include convenience and economics of operation. The basic tools for IGC are inexpensive, rugged, widely available, and as well suited for routine laboratory applications, as they are for demanding fundamental research. IGC data may be collected quite rapidly over extended temperature ranges. A variety of probes may be used in the mobile phase to elucidate the characteristics of the stationary phase, characteristics that otherwise are only obtained at far greater expenditure of time and money.

Perhaps a more important motivator for the increasing use of IGC is the method's flexibility and potential for generating data useful in the broad domain of polymer physical chemistry. Key to this consideration is the relationship between thermodynamics and the fundamental datum of IGC, the specific retention volume, V_g. The relationship has been

0097–6156/89/0391–0001$06.00/0
© 1989 American Chemical Society

discussed in detail elsewhere (4-8). In the present volume, Guillet and co-workers (9) demonstrate that the link between V_g and enthalpies of interaction and between the IGC datum and χ, the Flory-Huggins interaction parameter for polymer/probe pairs (10), is intrinsically valuable. Once again, IGC is a convenient route to information that is otherwise difficult to obtain and frequently unattainable. In all conventional experiments capable of measuring the interaction thermodynamics of polymers, such as swelling resulting from solvent uptake, neutron scattering, and changes in colligative properties, the polymer phase is highly diluted. In contrast, IGC measures interaction quantities at high polymer concentrations, thereby adding a considerable measure of practical relevance to IGC as a tool for materials science. Since a wide temperature range often is accessible, the important question of temperature dependence for χ can be resolved, at least in principle, without recourse to hypothesis or approximation.

The most vital element in the growth pattern of IGC is found in the breadth of applications. Aside from providing thermodynamic information, the IGC technique remains an excellent source for determining phase transitions, for measuring adsorption properties, and for estimating dispersive and non-dispersive forces acting at surfaces and interfaces. Given this spectrum of applications and the continued growth of materials science, it is reasonable to conclude that IGC methodology will continue to play an important role in furthering an understanding of materials behavior, and in helping to design multicomponent systems that meet desired targets of performance and durability.

Methodology and Instrumentation

IGC lends itself to automation and to computerized data processing, thereby enhancing the attractiveness of IGC as a laboratory technique. Guillet and co-workers report here on the automation of IGC temperature scanning, a measure certain to prove useful for routine determinations of retention data over broad temperature ranges (9). The application of algorithms to stimulate IGC behavior patterns by computational methods, as addressed here by Hattam, Du, and Munk (11), suggests that mathematical modeling of retention characteristics may develop along lines already well established in polymer processing. Consequently, for given combinations of stationary and volatile phases, adsorption, absorption, and diffusion patterns may be predicted and interpreted in terms of evolving concepts of component interaction.

A further significant trend in IGC technology is the variety of forms in which the stationary phase may be prepared. Elsewhere in this volume, Bolvari, Ward, Koning, and Sheehy (12) describe in detail two common methods for creating polymeric stationary phases. The most common methodology consists of depositing a thin polymeric coating on small, inert spheres and packing the column with the coated spheres. The second method, described in detail by Bolvari et al. (12), consists of depositing a thin polymeric coating on the inside wall of the column. Purportedly, the sensitivity and reliability of IGC can be improved by using the *capillary chromatography* approach. Matsuura and co-workers (13) use a simpler approach to creating a polymeric stationary phase; they use a finely divided and sieved polymer powder. This approach is similar to that used to study glass and carbon fiber stationary phases. Again, Bolvari et al. (12) describe in detail the methodology for creating a stationary phase of short fiber fragments as well as long, continuous fiber lengths.

Sorption and Diffusion in Polymers

Sorption and diffusion in polymers are of fundamental and practical concern. However, data acquisition by conventional methods is difficult and time consuming. Again, IGC represents an attractive alternative. Shiyao and co-workers, concerned with pervaporation processes, use IGC to study adsorption phenomena of single gases and binary mixtures of organic vapors on cellulosic and polyethersulfone membrane materials (13). Their work also notes certain limitations to IGC, which currently restrict its breadth of application. Notable is the upper limit to gas inlet pressure, currently in the vicinity of 100 kPa. Raising this limit would be beneficial to the pertinent use of IGC as an indicator of membrane-vapor interactions under conditions realistic for membrane separation processes.

Demertzis and Kontominas investigate the diffusion of water in polyvinylidene chloride polymers and copolymers, a subject of great importance in the area of packaging films (14). They note that most of the data on diffusion have been obtained by time consuming static methods requiring extensive data analysis. Although their work was limited to low concentrations of the diffusing molecules, the speed and convenience of obtaining diffusion coefficients and thermodynamic parameters relating to the sorption process are noteworthy features. Finally, Arnould and Laurence report on measurements of diffusion coefficients in polymethyl methacrylate of small species including methanol, acetone, alkyl acetates, and various aromatic hydrocarbons (15). The flexibility of IGC is demonstrated again by the breadth of the temperature range used in this work. The capillary IGC experiments lead to interpretations of the relationship between the size of the diffusing molecule and the diffusion coefficients, and provide a convenient data base for advancing diffusion theory for polymeric membranes.

Interaction characteristics in polymer-related areas frequently make use of solubility parameters (16). While the usefulness of solubility parameters is undeniable, there exists the limitation that they need to be estimated either by calculation or from indirect experimental measurements. The thermodynamic basis of IGC serves a most useful purpose in this respect by making possible a direct experimental determination of the solubility parameter and its dependence on temperature and composition variables. Price (17) uses IGC for the measurement of accurate χ values for macromolecule/vapor pairs, which are then used for the evaluation of solubility parameters for a series of non-volatile hydrocarbons, alkyl phthalates, and pyrrolidones. It may be argued that IGC is the only unequivocal, experimental route to polymer solubility parameters, and that its application in this regard may further enhance the practical value of that parameter. Guillet (9) also notes the value of IGC in this regard.

Polymer Blend Characterization

The suitability of IGC as a route to interaction thermodynamics using non-volatile stationary phases and selected probe molecules at high dilution has been noted above. Much valuable information on the miscibility of solvent-polymer systems, derived from IGC measurements, continues to be published in the literature. However, equally important is information on the state of interaction among the non-volatile components of complex polymer-containing systems. Such information is an invaluable guide to the formulation of polymer blends and fiber- and particulate-reinforced polymer compounds, and would appear to have at least equal relevance to the properties of high performance, non-polymeric composite materials. Other important and rapidly-growing areas of science, such as bioengineering, also would be well served by a convenient experimental method for the thermodynamic characterization of relevant materials.

In theory, IGC is well suited for the study of mixed polymer systems (18-22) and must be considered along with traditional methods for measuring polymer-polymer interactions (23), such as melting point depression (24), heat of mixing (25-27), cloud point (28-30), light scattering (31), osmotic pressure (32), and interdiffusion via forward recoil spectrometry (33-35). By using IGC, polymer mixtures may be studied over the entire composition range and at all accessible temperatures. However, in practice it is found that the interaction parameter for a pair of stationary phase components, generally written χ_{23}, is not uniquely defined for a given polymer composition and temperature. Rather, it varies with the selection of probe, thereby creating a dilemma that is yet to be resolved fully. Conceptually, such variations should not be entirely surprising. IGC tests the interaction thermodynamics at polymer/polymer contacts by injecting a small amount of sensing or probe molecules. Unless the volatile phase molecule partitions randomly between the components of the stationary phase, some perturbation in the energies at polymer/polymer contacts should be expected. Indeed, it might be argued that the thermodynamic system is changed by each incidence of non-random partitioning. The measured value of χ_{23} would then be valid for each specific system as defined by the polymer mixture and the probe. However, it should not be regarded as an unequivocal measure of the thermodynamic state pertaining to the polymers alone.

The importance of applying IGC to the interaction thermodynamics of polymers is well illustrated by the content of the present volume. While a general solution to the probe dependence problem may not yet be available, what may be called interim approaches are followed by diPaola-Baranyi (36) and by Klotz and co-workers (37). Here, the probe-to-probe variations of χ_{23} in the system polystyrene/poly(2,6-dimethyl-1,4-phenylene oxide), (PS/PMMPO) are acknowledged. Both authors found the variations to be non-systematic, therefore justifying a simple averaging procedure. Significant differences are found in the interaction numbers reported by these authors, the averaging process notwithstanding. The miscibility question is also investigated for other polymer blends by diPaola-Baranyi, again relying on an averaging procedure for the calculation of interaction parameters.

Munk and co-workers have been concerned with the above-stated problem for some time (38, 39). In this volume (40), their attention is focused on miscible blends of polycaprolactone and polyepichlorohydrin. These authors demonstrate that to a considerable degree the probe variation problem can be mitigated by scrupulous attention to experimental details in the IGC methodology. This concern for details is required at any rate, if the high data reproducibility needed for meaningful studies of interaction in miscible polymer blends is to be attained. These details center on modified methods for coating polymers onto solid supports, on improved methods for measuring carrier gas flow rates, and on enhanced, computer-based data analyses of elution traces. Also, corrections are made for contributions to retention times from uncoated support material. More than twenty volatile probes are used by Munk, and the probe-to-probe variations in χ_{23}, while not entirely absent, are much less apparent than they would be under standard experimental protocols.

Since relatively slight variations in interaction parameter values can cause significant shifts in the degree of component miscibility, the demand for high accuracy in IGC measurements is paramount. Su and Fried (41) have applied the modifications first suggested by Munk (38, 39) to their work on blends using polystyrene, poly(4-methylstyrene), and PMMPO. They found evidence for thermal degradation in PMMPO when columns of this polymer were exposed repeatedly to temperatures in the span 200 to 280°C. Therefore, they identify another potential source of difficulty to be heeded if the true potential of the IGC route to interaction thermodynamics in polymer mixtures is to be realized.

Clearly, the use of IGC to generate formal thermodynamic information brings into play, first, the method's great convenience and flexibility, and second, the limitations imposed by the cited volatile-phase dependencies. It is likely that a full resolution of the problem hinges on a more rigorous definition of the thermodynamic terms that pertain to binary polymer systems. According to Sanchez (42), when compositional dependencies are encountered, then a full description of polymer mixture thermodynamics requires the definition of four different χ parameters. One χ is associated with the free energy, two are related to the first concentration derivative of the free energy, and the fourth χ is related to the second concentration derivative. The procedure needed to obtain these parameters involves using appropriate equation-of-state models and theoretically derived Henry's law constants for the process of gases sorbing on polymer solids. Sanchez (42) derives such constants and suggests that when applied to IGC, their use will produce bare interaction parameters, independent of mobile phase composition. Developments along these, or related lines, will provide further impetus to the important task of clarifying the thermodynamic criteria of interaction in multicomponent polymer systems, and to the important part to be played in that task by IGC.

Surface and Interface Characterization

Because of the current emphasis on high performance reinforced polymer composites, much attention is being placed on fiber-reinforced polymer matrices as subjects of study. This attests to the great importance of the interface and interphase in determining the properties of such systems, and on the relatively sparse information currently available on the subject. The concept of acid/base interactions across the fiber-polymer interface is noted particularly. The relevance of acid/base theories to the behavior of polymers at surfaces and at interfaces has been studied by Fowkes (43), among others, using laborious calorimetric measurements of interaction enthalpies (43). Once again, data acquisition via IGC appears to be sufficiently rapid and accurate to have generated appreciable advances. A good illustration of IGC's pertinence to the matter is documented by Schultz and Lavielle, who use dispersion force probes along with volatiles known to act as Lewis acids or bases, to evaluate the dispersive and non-dispersive force contributions to the surface energies of variously surface-treated carbon fibers (44). They use the Gutmann theory (45) to obtain acceptor and donor numbers for their substrates, as well as for an epoxy matrix. The adhesion of the fiber-matrix interface depends clearly on the measured strength of acid/base interactions.

Carbon fiber reinforced composites are at the forefront of current developments in polymer composites, and there is additional evidence for the important role being played by IGC in characterizing the interface in such systems. The Gutmann theory is used by Bolvari and Ward, who report acid/base interactions for surface-treated carbon fibers and a series of thermoplastic polymer hosts, including polysulfone, polycarbonate, and polyetherimide (46). Once again, strong acid/base coupling is found to be beneficial to the strength of the interface. Wesson and Alfred investigate carbon fibers (47) and compare the surface properties of graphitized carbon fibers with sized versions and with fibers treated in radio-frequency glow discharges (plasmas). In addition to demonstrating the effects of surface modification procedures on surface acid/base character, the IGC technique is used to produce adsorption isotherms for the fiber substrates. In this manner, site energy distributions are obtained that emphasize differences between the uniform surface energetics of the graphitized fiber and the sized or plasma-treated versions.

Since IGC is able to generate adsorption isotherms and to evaluate acid/base interactions for specified adsorbate-adsorbent pairs, it follows that the technique is able to develop a detailed picture of surface properties for non-volatile stationary phases. This is illustrated, again for carbon fibers, by Vukov and Gray (48). They combine IGC information at essentially zero coverage of the injected probes with finite concentration data to obtain heat of adsorption values ranging from zero to multi-layer coverage. Their meticulous study shows the effects of thermal pretreatment on fiber surface characteristics, and underscores the convenience and power of IGC to generate information otherwise far more difficult to obtain.

A further illustration of IGC as a source of data for acid/base characterization of polymers and of solid constituents of complex polymer systems, is given by Osmont and Schreiber (49), who rate the inherent acid/base interaction potentials of glass fiber surfaces and of polymers by a comparative index, based on the Drago acid/base concepts (50). The interaction index is conveniently measured by IGC and is shown to differentiate clearly among untreated and variously silane-modified glass fiber surfaces. Conventional methods are used to determine adsorption isotherms for fiber-polymer pairs, and the IGC data are used to demonstrate the relationship between acid/base interactions and the quantity of polymer retained at fiber surfaces.

The applicability of IGC to particulates, used as pigmenting or reinforcing solids in polymer matrices, has been noted above. In surface coatings, pigment-polymer interaction may strongly affect adhesion, mechanical integrity, and durability of protective polymer films. The use of IGC on particulate substrates is illustrated in this volume by Papirer and co-workers (51). They characterize the surface properties of high surface area silicas both as supplied by manufacturers and as surface modified by grafting to them alkyl, diol, and polyethylene glycol moieties. The grafting procedures are shown to lead to important changes in donor-acceptor properties and consequently to the suitability of these particulates as reinforcing materials for polymer or elastomer matrices. Papirer's work also demonstrates the feasibility of relating surface characteristics obtained by IGC with independent surface analyses produced by nuclear magnetic resonance and X-ray photoelectron spectroscopy. The correlations attest to the validity of IGC techniques as surface diagnostic tools.

Guillet (9) uses IGC to estimate the degree of crystallinity in semicrystalline polymers and to compute the surface area of polymer powders. Linear polyethylene is used as the vehicle to demonstrate the former application. In the latter, a requisite is to evaluate the partition coefficient for a selected probe/polymer combination (n-decane/PMMA in the present instance). Once this is obtained via IGC, simple retention time measurements become suitable as routine analytical or control methods to monitor surface areas in polymer powders.

Analytical Applications

Laub and Tyagi investigate the analytical qualities of GC in general, and IGC specifically (52). They demonstrate the value of family retention plots from which the elution behavior of homologues or of related compounds in a series of volatile phase components may be estimated. In this way, the separability of such compounds via IGC methods is readily predicted. Also noted is the economy of sample sizes required for IGC, an invaluable consideration when only minute quantities of material are available.

One variant of IGC is pulsed chromatography, which allows for a monitoring of changes brought about in a stationary phase by chemical, environmental responses, and the like. Raymer and co-workers use deuterated tracers in pulsed chromatography to study the sorption of polar and non-polar probes on various imide-based polymers (53). A specific aspect of the study centered on the influence of water on the retention characteristics of given polymer/probe pairs. Emphasis is placed on the potential value of the technique in determining break-through volumes of specific molecules for selected polymer barrier structures.

Special Applications

Neill and Winans (54) and Gilbert (55) demonstrate the applicability of IGC to systems other than synthetic polymers and fibers. These workers have expanded the use of IGC to include research involving naturally-occurring materials. Using a column packed with a mixture of finely divided coal and non-porous glass beads, Neill and Winans (54) utilize IGC to follow the chemical and physical changes that occur when coal is heated in inert atmospheres. They are able to observe differences in transition temperatures and enthalpies of sorption for the different coals studied. Gilbert (55) applies a modified frontal analysis method to study water sorption kinetics in biological macromolecules. By doing so, Gilbert avoids having to apply equilibrium assumptions to these systems, which are influenced by entropic as well as enthalpic considerations.

Conclusion

This overview outlines some of the important basic concepts implicated in IGC, notes some of the strengths and limitations inherent in the technique, and mentions at least the more active areas of application for IGC. In a field as fertile and as rapidly changing as IGC, comprehensiveness would entail lengthy discussion, and ultimately would fail to account for all that is noteworthy. An attempt has been made to stress the breadth of possible applications for IGC, and this supports the tenet that the method will continue to play an expanding role in the science and technology of polymers and of advanced materials in general. It is regrettable that a book of the present size and scope can sample only a small fraction of the total output in IGC. One may look forward with confidence to the further evolution of a methodology at once subtle yet simple and convenient, at once rigorous in its thermodynamic basis and useful for analytic or quality-control objectives. The growth rate in IGC, alluded to at the beginning of this overview, will no doubt be maintained in the future. If that projection is correct, then future volumes of this kind will no doubt follow.

Literature Cited

1. Kiselev, A.V. In Advances in Chromatography; Giddings, J.C.; Keller, R.A., Eds.; Marcel Dekker Co.: New York, 1967.
2. Smidsrod, O.; Guillet, J.E. Macromolecules 1976, 2, 272.
3. Laub, R.J. Private communication, and article to be published.
4. Braun, J.M.; Guillet, J.E. Adv. Polym. Sci. 1976, 21, 108.
5. Gray, D.G. Proc. Polym. Sci. 1977, 5, 1.
6. Laub, R.J.; Pecsok, R.L. Physicochemical Applications of Gas Chromatography; Wiley and Sons: New York, 1978.
7. Conder, J.R.; Young, C.L. Physicochemical Measurements by Gas Chromatography; Wiley and Sons: New York, 1979.
8. Lipson, J.E.G.; Guillet, J.E. In Development in Polymer Characterization; Dawkins, J.V., Ed.; Applied Science Publ.: London, 1982; Vol. 3, Chapter 2.
9. Guillet, J.E.; Romansky, M.; Price, G.J.; van der Mark, R. In Inverse Gas Chromatography; Lloyd, D.R.; Schreiber, H.P.; Ward, T.C., Eds.; ACS Symposium Series No. 391; American Chemical Society: Washington, D.C., 1989.
10. Flory, P.J. Principles of Polymer Chemistry; Cornell University Press: Ithaca, N.Y., 1953; Chapter XIII.
11. Hattam, P.; Du, Q; Munk, P. In Inverse Gas Chromatography; Lloyd, D.R.; Schreiber, H.P.; Ward, T.C., Eds.; ACS Symposium Series No. 391; American Chemical Society: Washington, D.C., 1989.
12. Bolvari, A.E.; Ward, T.C.; Koning, P.A.; Sheehy, D.P. In Inverse Gas Chromatography; Lloyd, D.R.; Schreiber, H.P.; Ward, T.C., Eds.; ACS Symposium Series No. 391; American Chemical Society: Washington, D.C., 1989.
13. Shiyao, B.; Sourirajan, S.; Talbot, F.D.F.; Matsuura, T. In Inverse Gas Chromatography; Lloyd, D.R.; Schreiber, H.P.; Ward, T.C., Eds.; ACS Symposium Series No. 391; American Chemical Society: Washington, D.C., 1989.
14. Demertzis, P.G.; Kontominas, M.G. In Inverse Gas Chromatography; Lloyd, D.R.; Schreiber, H.P.; Ward, T.C., Eds.; ACS Symposium Series No. 391; American Chemical Society: Washington, D.C., 1989.
15. Arnould, D.; Laurence, R.L. In Inverse Gas Chromatography; Lloyd, D.R.; Schreiber, H.P.; Ward, T.C., Eds.; ACS Symposium Series No. 391; American Chemical Society: Washington, D.C., 1989.
16. Barton, A.F.M. Handbook of Solubility Parameters and Other Cohesion Parameters. CRC Press: Boca Raton, Florida, 1983.
17. Price, G.J. In Inverse Gas Chromatography; Lloyd, D.R.; Schreiber, H.P.; Ward, T.C., Eds.; ACS Symposium Series No. 391; American Chemical Society: Washington, D.C., 1989.
18. Deshpande, D.D.; Patterson, D.; Schreiber, H.P. Macromolecules 1974, 7, 630.
19. Olabisi, O. Macromolecules 1975, 8, 316.
20. Robard, A.; Patterson, D. Macromolecules 1977, 10, 1021.
21. Su, C.S.; Patterson, D.; Schreiber, H.P. J. Appl. Polym. Sci. 1976, 20, 1025.
22. Walsh, D.; McKeon, J.G. Polymer 1980, 21, 1335.
23. Paul, D.R.; Barlow, J.W.; Keskkula, H. In Encyclopedia of Polymer Sciences and Engineering, Second Edition; Wiley & Sons, Inc.: New York, 1988; Volume 12, Page 399.
24. Nishi, T.; Wang, T.T. Macromolecules 1975, 6, 909.

25. Weeks, N.E.; Karasz, F.E.; MacKnight, W.J. J. Appl. Phys. 1977, 48, 4068.
26. Barlow, J.W.; Paul, D.R. Polym. Eng. Sci. 1987, 27, 1482.
27. Uriarte, C.; Eguiazabal, J.I.; Llanos, M.; Iribarren, J.I.; Iruin, J.J. Macromolecules 1987, 20, 3038.
28. Ueda, H.; Karasz, F.E. Macromolecules 1985, 18, 2719.
29. Balazs, A.C.; Karasz, F.E.; MacKnight, W.J.; Ueda, H.; Sanchez, I.C. Macromolecules 1985, 18, 2784.
30. Ronca, G.; Russell, T.P. Macromolecules 1985, 18, 665.
31. Fukuda, T.; Nagata, M.; Inagaki, H. Macromolecules 1984, 17, 548.
32. Shiomi, T.; Kohno, K.; Yoneda, K.; Tomita, T.; Miya, M.; Imai, K. Macromolecules 1985, 18, 414.
33. Green, P.F.; Doyle, B.L. Macromolecules 1987, 20, 2471.
34. Composto, R.J.; Kramer, E.J. Polymer Preprints 1988, 29, 401.
35. Composto, R.J.; Kramer, E.J.; White, D.M. Macromolecules In press, 1988.
36. diPaola-Baranyi, G. In Inverse Gas Chromatography; Lloyd, D.R.; Schreiber, H.P.; Ward, T.C., Eds.; ACS Symposium Series No. 391; American Chemical Society: Washington, D.C., 1989.
37. Klotz, S.; Gräter, H.; Cantow, H.-J. In Inverse Gas Chromatography; Lloyd, D.R.; Schreiber, H.P.; Ward, T.C., Eds.; ACS Symposium Series No. 391; American Chemical Society: Washington, D.C., 1989.
38. Al-Saigh, Z.Y.; Munk, P. Macromolecules 1984, 17, 803.
39. Card, T.W.; Al-Saigh, Z.Y.; Munk, P. J. Chrom. 1985, 301, 261.
40. El-Hibri, M.J.; Cheng, W.; Hattam, P.; Munk, P. In Inverse Gas Chromatography; Lloyd, D.R.; Schreiber, H.P.; Ward, T.C., Eds.; ACS Symposium Series No. 391; American Chemical Society: Washington, D.C., 1989.
41. Su, A.C.; Fried, J.R. In Inverse Gas Chromatography; Lloyd, D.R.; Schreiber, H.P.; Ward, T.C., Eds.; ACS Symposium Series No. 391; American Chemical Society: Washington, D.C., 1989.
42. Sanchez, I.C. Polymer In press, 1988.
43. Fowkes, F.M. J. Adhesion Sci. Tech. 1987, 1, 7.
44. Schultz, J.; Lavielle, L. In Inverse Gas Chromatography; Lloyd, D.R.; Schreiber, H.P.; Ward, T.C., Eds.; ACS Symposium Series No. 391; American Chemical Society: Washington, D.C., 1989.
45. Gutmann, V. The Donor-Acceptor Approach to Molecular Interactions. Plenum Press: New York, 1983.
46. Bolvari, A.E.; Ward, T.C. In Inverse Gas Chromatography; Lloyd, D.R.; Schreiber, H.P.; Ward, T.C., Eds.; ACS Symposium Series No. 391; American Chemical Society: Washington, D.C., 1989.
47. Wesson, S.P.; Alfred, R.E. In Inverse Gas Chromatography; Lloyd, D.R.; Schreiber, H.P.; Ward, T.C., Eds.; ACS Symposium Series No. 391; American Chemical Society: Washington, D.C., 1989.
48. Vukov, A.; Gray, D.G. In Inverse Gas Chromatography; Lloyd, D.R.; Schreiber, H.P.; Ward. T.C., Eds.; ACS Symposium Series No. 391; American Chemical Society: Washington, D.C., 1989.
49. Osmont, E.; Schreiber, H.P. In Inverse Gas Chromatography; Lloyd, D.R.; Schreiber, H.P.; Ward, T.C., Eds.; ACS Symposium Series No. 391; American Chemical Society: Washington, D.C., 1989.
50. Drago, R.S.; Vogel, G.C.; Needham, T.E. J. Amer. Chem. Soc. 1971, 93, 6014.

51. Papirer, E.; Vidal, A.; Balard, H. In Inverse Gas Chromatography; Lloyd, D.R.; Schreiber, H.P.; Ward, T.C., Eds.; ACS Symposium Series No. 391; American Chemical Society: Washington, D.C., 1989.

52. Laub, R.J.; Tyagi, O.S. In Inverse Gas Chromatography; Lloyd, D.R.; Schreiber, H.P.; Ward, T.C., Eds.; ACS Symposium Series No. 391; American Chemical Society: Washington, D.C., 1989.

53. Raymer, J.H.; Cooper, S.D.; Pellizzari, E.D. In Inverse Gas Chromatography; Lloyd, D.R.; Schreiber, H.P.; Ward, T.C., Eds.; ACS Symposium Series No. 391; American Chemical Society: Washington, D.C., 1989.

54. Neill, P.H.; Winans, R.E. In Inverse Gas Chromatography; Lloyd, D.R.; Schreiber, H.P.; Ward, T.C., Eds.; ACS Symposium Series No. 391; American Chemical Society: Washington, D.C., 1989.

55. Gilbert, S.G. In Inverse Gas Chromatography; Lloyd, D.R.; Schreiber, H.P.; Ward, T.C., Eds.; ACS Symposium Series No. 391; American Chemical Society: Washington, D.C., 1989.

RECEIVED November 15, 1988

METHODOLOGY AND INSTRUMENTATION

Chapter 2

Experimental Techniques for Inverse Gas Chromatography

A. E. Bolvari, Thomas Carl Ward[1], P. A. Koning[2], and D. P. Sheehy[3]

Department of Chemistry, Polymer Materials and Interfaces Laboratory, Virginia Polytechnic Institute and State University, Blacksburg, VA 24061

Details for producing optimum performing packed, capillary, and fiberfilled columns in inverse gas chromatography experiments are discussed. Also, the crucial factors that might lead to instrumental error in this technique are evaluated and cautions are provided.

In inverse gas chromatography (IGC), the interactions of gaseous probe molecules with a stationary phase contained within a column results in a characteristic retention time, t_p, which can be translated into a number of important thermodynamic, kinetic, and surface properties. The theory and principles have been well developed and reviewed in other sources ($\underline{1}$). However, to obtain meaningful data, one must design and perform experiments with an awareness of the assumptions and limitations of both the theory and the measurements. There are many small numerical corrections and tedious technical requirements for success that individually seem insignificant; however, when taken as a whole, they can determine the success or failure of the research. A careful analysis of each of the areas of application, for example, diffusion constants, solubility parameters, and activity coefficients, reveals that the cumulative nature of the IGC inaccuracies makes careful technique imperative if meaningful absolute results are to be obtained. Examination of Equations 1,2, and 3 reveal the origin of some of the possible errors that may enter the data. These are elaborated in the following text.

The key parameter in the IGC measurements is the specific retention volume, V_g^o, or the amount of carrier gas required to elute a probe from a column containing one gram of <u>interacting</u> stationary phase material. The exact quantity in terms of experimental variables is:

[1]Address correspondence to this author.
[2]Current address: Amoco Chemicals Corporation, Naperville, IL 60566
[3]Current address: 3M Company, St. Paul, MN 55144

0097–6156/89/0391–0012$06.00/0

$$V_g^O = \frac{273.2}{T} \cdot \frac{(t_p - t_o)}{W} \cdot F \cdot J \cdot C \qquad (1)$$

where

$$C = 1 - \left(\frac{P_{H_2O}}{P_o}\right) \qquad (2)$$

and

$$J = 1.5\left[\frac{(P_i/P_o)^2 - 1}{(P_i/P_o)^3 - 1}\right] \qquad (3)$$

In these equations, T is the temperature of the flowmeter, which measures the carrier gas flow rate F; W is the weight of polymer on the column; t_p is the retention time of the probe molecule; t_o is the retention time of a noninteracting marker such as air; J is the correction for the pressure drop across the column; C is the correction for the vapor pressure of water in the soap bubble flowmeter; P_o is the pressure of the carrier gas at the outlet of the column (atmospheric pressure); P_i is the pressure at the inlet; and PH_2O is the vapor pressure of water at the temperature of the flowmeter. Thus, the pressures, times, weights, and flow rates are critical for precise calculations and are examined in this paper. First, it will be useful to discuss the types of stationary phases before proceeding with the details of proper experimental procedure.

Inverse gas chromatography has proven to be a particularly important technique for the investigation of polymers, with most studies making use of packed columns. IGC also has been recently extended to the investigation of fibers and of polymers coated on capillary columns. The preparation of each of these columns is very important to overall success.

Indeed, the most essential piece of equipment in IGC is the chromatographic column. The function of the column is to encourage repetitive partitioning of each solute molecule between the gas and the liquid or solid phase under conditions that minimize the range of retention times exhibited by identical molecules of each solute (2). The accuracy of the results is directly influenced by the degree to which this goal has been achieved. In this regard, retention diagrams, in which the natural log of the specific retention volume is plotted versus reciprocal absolute temperature, are excellent indicators of overall column performance (1). Because of the ease of temperature control in a gas chromatograph, one may explore the polymer using the probe above and below various possible phase transitions, with a linear response being revealed by the retention diagram when operating under equilibrium conditions.

Packed Columns

Packed columns are suitable for a wide range of investigations, including low molecular weight materials, homo- and copolymers, blends, and block copolymers (3). In these investigations, the

stationary phase consists of a noninteracting, finely divided solid
that was coated with a thin film of the polymer or polymer blend.
To achieve this, the polymer is dissolved in the solvent of choice
(desirably of high vapor pressure) and filtered. A quantity of
inert support is weighed to achieve a 15 wt/wt-% stationary phase.
An acid-washed, dimethylchlorosilane treated support is usually
chosen. One widely used support is Chromasorb W, manufactured by
Johns Mansville. Alternatively, 60/80 mesh sized glass beads may
be used. Successful loadings range in the 4 to 13 wt-% category
for Chromasorb W and at approximately 0.5 wt-% for glass beads.
The inert support and filtered polymer solution are placed into a
hedgehog flask. The hedgehog flask is a 500 mL pear shaped flask
with vigreux fingers extending to its interior. Coating is
achieved by slow evaporation of the solvent using a rotary
evaporator. The rotary motion, in combination with the hedgehog
design, assures gentle but adequate agitation needed for complete
and uniform coverage of the support. Al-Saigh and Munk have
reported a novel soaking technique, requiring several hours of
work, that reportedly minimizes loss of polymer and aids in
determining packing loadings (4). The coated support must be then
dried in a vacuum oven. In order to ensure maximum solvent
removal, the oven temperature should be adjusted to just below the
polymer glass transition temperature and drying maintained for 48
hours. Following this treatment, the packing becomes a free
flowing powder if the film is thin enough, even for low molecular
weight waxes. The support is sifted through a 60/80 mesh sieve to
ensure even particle sizes. Any large agglomerates and fines
(fractured solid support) that are not removed in this step would
reduce packing efficiency, cause peak tailing, and expose uncoated,
non-neutralized surfaces.

     To begin packing the column, stainless steel tubing is
straightened (1 to 2 m), silane treated steel wool is used to plug
one end, and this same end is attached to a water aspirator. The
container holding the stationary phase is weighed. A few grams are
put in the columns and packing is accomplished with the aid of a
mechanical vibrator. The supported stationary phase is continually
added and packed until the column is filled, at which time the
container is weighed again. These weighings must be as precise as
can be achieved so that the exact amount of stationary phase in the
column is known. The other end of the column is sealed with glass
wool; the column is coiled and placed in the chromatograph with
only its inlet port connected. After a 30 minute purge with helium
gas, the column is taken to 10 degrees above the glass transition
temperature (Tg) or the crystalline melting point (Tm), whichever
is highest, of the polymer(s) and held at this temperature for 12
hours. This further allows for removal of any residual solvent and
solvent-induced morphologies present in the polymer. Complete
solvent removal is essential. In any experiment that is
subsequently conducted and shows a time-dependent or
nonreproducible character, residual solvent should be suspected.

     The weight of the polymer on the column is determined most
commonly by ashing, using a thermogravimetric analysis system
(TGA). A typical experiment requires 10 mg of coated support.

This is loaded onto a TGA sample pan and then the TGA oven temperature is raised to 550°C in an oxygen atmosphere. Measurements are made both on the coated and uncoated support to correct for weight loss of the inert solid support. Since almost all calculations in IGC require knowledge of the specific retention volume, an accurate determination of the amount of coating is crucial. This is generally regarded as one of the most likely sources of error in the IGC experiment.

Certain polymers can not be successfully burned off of the support and alternative methods must be used to determine loading. For example, soxhlet extractions must be used for most siloxane containing polymers. Multiple specimens must be tested to get the error in loading to less than 1% for most calculations.

The amount of loading found by the described methods is typically 3% lower than that expected from the initial weight of polymer and support. This is due to coating of the glass flask with the polymer. The optimum loading lies between 6 to 15 wt/wt-%. Several researchers (5,6) have shown that below 4 to 6 wt/wt-% loading, a contribution to retention times due to adsorption on uncoated support can be detected, while at higher loadings (<15 wt/wt-%), diffusion through the thicker polymer coatings becomes a problem.

Fiber Columns. Columns are constructed from 1/4 inch stainless steel tubing with passivated inner walls. Approximately 1.0 m lengths of stainless steel are cut from the coil, fitted with swagelok nuts and ferrules, and then weighed. The stainless steel is straightened for loading. Packing the fibers into the column will vary with the physical nature of the fiber. For example, short, chopped fibers are best reduced to lengths that pass through a fine mesh and then pumped or vibrated into the tubing. However, continuous filaments may be aligned into tows, pulled into place, and the ends trimmed to fit the column. In the case of carbon fibers of approximately 5 μm diameter, approximately six 1.2 m long loops or tows of the fiber are attached to a wire, which is inserted in the column and drawn through it. This loads ca. 150,000 individual strands. Obviously, the goal in all cases is to provide maximum possible fiber surface area. The column is then weighed again for determination of the sample weight. It is a good idea if the column is conditioned at 110°C for 12 hours to remove water.

It should be obvious that the packed fiber column can be regarded as a chemical reactor for fiber surface modification, involving either gas or liquid phase chemistries. Thus, a powerful and convenient system for exploring surface properties is created by careful construction of the initial column.

Capillary Columns. Capillary columns are long, open tubes of small diameter. They have high efficiencies, low sample capacity, and low pressure drop. Commercially available capillary columns range from 0.1 to 0.53 mm in internal diameter and from 5 to 50 m in length. The inside wall of the tubing is coated with a film ranging from 0.1 μm to thick films of 3.0 μm.

A major advantage of capillary columns is that many total chromatographic plates are obtainable. Plates per meter of column

length are comparable with packed columns, but much longer columns
are usable since capillary columns have extremely low pressure
drops. Capillary columns have been made from stainless steel,
glass, and fused silica. The brittleness of glass is a major
disadvantage that has been circumvented through the use of fused
silica columns. As much as 60 m of fused silica has proven
successful (7).

Glass has been preferred as a column material because of its
more inert character. Fused silica is a high purity glass that is
composed of essentially silicon and oxygen. It is the type of
glass that is most inert, has the best flexibility, and produces
the most uniform product that responds most predictably to
subsequent coating (8). Fused silica is therefore the material of
choice for capillary IGC experiments.

In preparing capillary columns, the primary goal should be the
deposition of a uniform film throughout the column. Coating
techniques can be fitted into one of two general methods, one is
termed "dynamic" and the other the "static" technique. The dynamic
method requires 10 mL of dilute polymer solution (6 to 10 wt-%) to
be placed in a reservoir and pushed through the column with
nitrogen at approximately 0.5 atm pressure. Continued N_2 flow
dries the polymer, which adheres to the tube's inner walls. A 10
mL portion of the coating solution is put through the column as
many times as it takes to obtain the desired film thickness. The
static technique is the one currently used by most researchers and
is the one that is elaborated below.

The column is filled with a coating solution whose
concentration will determine the film thickness. It is important
that the solution be dust free and degassed to eliminate bumping
during the solvent evaporation step (low-boiling solvents are
preferred). This may be achieved by filtering a half-strength
solution and then boiling this to half volume to accomplish
degassing (9). The covered solution is cooled rapidly and the
column is filled. Others prefer to degas the coating solution by
subjecting it to an ultrasonic treatment (7). Once the column has
been filled, sufficient additional liquid is drawn through the
column to eliminate axial concentration gradients that may have
formed while the column was being filled. One end of the column is
then sealed with a commercial epoxy. After the seal has hardened,
the column is put in a constant temperature bath at 35°C and
connected to a vacuum system via the open end. The coating
solution is evaporated under partial vacuum for the first few hours
to suppress spontaneous boiling. After that time, full vacuum is
applied and the coating solution evaporates at a slow steady rate.
The drying rate is dependent upon the polymer, solvent, and
solution concentration. Typically several days may be required.
Again, the importance of removing all solvent as completely as
possible can not be overemphasized.

The weight of the polymer on capillary columns is obtained by
rinsing the columns with approximately 30 mL of the coating
solvent. The eluent is collected in a pre-weighed beaker. After
evaporation of the solvent, the beaker is weighed again. The
comments with respect to accuracy in this step, which were made
above for packed columns, apply in this case as well.

The small internal volume and thin liquid film of capillary columns require injection of small samples. If such small samples could be injected, the internal volume of the typical inlet system would probably be too large and would cause peak broadening. Also, it is not practical to reproducibly inject the small samples required. For these reasons, an inlet splitter is necessary. The sample is injected into a conventional septum inlet port. The sample is vaporized, mixed with the carrier gas and sent to the splitter. The sample is then split into two streams, one going to the capillary column and the other being vented to the atmosphere by means of an adjustable vent. The practice is to inject samples on the order of 1 to 2 µL and split the sample approximately 100 to 1. To eliminate or minimize band broadening after the column, the practice is to add a makeup gas to increase the linear velocity and decrease the residence time of the components as they are being swept into the detector. If the capillary column uses a flame ionization detector, then hydrogen gas serves the dual purpose of fuel and makeup gas. The makeup gas also allows optimum operating conditions of the flame ionization detector.

When capillary column temperature is raised, as will be necessary in the determination of enthalpies and entropies of probe/polymer interactions, the retention times of the probes will increase. This might seem odd since it is normal to expect an increase in temperature to result in a decrease in retention time. This behavior is due to the gas viscosity. When the temperature of a gas is increased, its viscosity is also increased (as opposed to liquids where the opposite is true). In a system having a constant pressure drop (as with open tubular columns), an increase in the viscosity results in a simultaneous decrease in the velocity of the carrier gas as shown in the following relationship:

$$\eta = p \, r_c^2 / 8 \, L \, \bar{u} \quad , \tag{4}$$

where η is the carrier gas viscosity at room temperature; p is the pressure drop; r_c is the column tube radius; L is the column length; and u is the average linear flow rate. For this reason, the flow rate must be adjusted at each temperature to compensate for the changing viscosity and hence the changing linear velocity of the carrier gas.

Instrumentation Considerations

Instrumentation for IGC has been fairly standard. Detectors should be chosen to most accurately reveal the probe molecule. Flame ionization is most common. Carrier gases are usually helium or hydrogen. It should be noted that the compressibility of the gas is always corrected for in packed column work because of the pressure drop across the column, as shown in Equations 1 to 3; but, this is negligible in capillary investigations.

Soap bubble flowmeters are commonly used to find gas velocities, but again are known to be one source of error that accumulates. Flow rates are usually measured from the column end using the soap bubble flowmeter modified with an inverted U-tube on top. The U-tube allows the flowmeter to be purged with the carrier

gas prior to measurement. This is important since diffusion of
helium or hydrogen through a bubble to the air-rich side will cause
errors in measurement at the slow flow rates (5 to 10 mL/min.) of
IGC. The walls of the flowmeter should be thoroughly dampened by
passing 20 bubbles across its entire length before any data are
taken. An average of four readings are needed to assure accuracy.
Pressures at the inlet of the column may be measured with a mercury
manometer. The outlet pressure (atmospheric) must be accurately
measured using a mercury and brass barometer applying the
appropriate corrections. Oven and ambient temperatures are
typically measured with a Pt resistance thermometer.

To measure the retention time, a small amount of the probe is
injected with a 10 μL syringe into the chromatograph along with air
or methane as a marker. Retention times are noted. Then traces of
the probe remaining in the syringe from the previous injection are
again introduced with air or marker into the column and the
retention time observed. The procedure is repeated until
consecutive injections show no dependence of the retention volume
on the amount of probe in the column. Typically this occurs when
the probe peak area is of the same order of magnitude as the 5 to
10 μLof air injected with the probe. The following injection
technique may be adopted in order to ensure the injection of the
desired infinitesimal sample size. Initially, the syringe is
flushed out many times with solvent vapor. Approximately 1 μL of
solvent vapor with 1 μL of a 0.08 vol-% mixture of methane in
helium is injected into the chromatograph and its response
recorded. Further, 1 μL injection of the methane/helium mixture is
made until the peak for the residual solvent vapor is no longer
recorded. This allows a regression of many sample sizes (via the
peak area) on the retention time to assure they are independent
(that is, slope = 0). Small, but significant, contributions to the
specific retention volume will exist due to interaction with the
support (10). These must be measured and subtracted from values
observed on loaded columns.

With the advent of microcomputers and associated
instrumentation, automation of IGC techniques is now possible.
Guillet (11) has used an automatic system for sample injection and
measurement. Sample injection can also be automated via a
commercially available headspace sampler (7). Its operation can
best be described by the following sequence. The first step is
pressurization of the sample vial using the carrier gas. The
second step is the filling of the sample loop. Vial vapor flows
through the sample loop as the pressure is permitted to drop toward
atmospheric. The last step is injection at which time the loop
contents are driven into the IGC injection port. After a pre-
selected time, the system returns to standby mode where a small
flow of carrier is purging the loop. The advantages of an
automated system include simplicity, convenience, and time saved in
running the sample. Perhaps the most significant gain, however,
from automated headspace sampling is reproducibility in retention
time measurements. For any thermodynamic calculations (for
example, Flory-Huggins Chi parameter), where a premium must be
placed on precision, serious investigators should strongly consider
this addition to their equipment.

Conclusions

The selection of injection mode and detection, options as to the
type and length of column, and the choice of carrier gas and
carrier gas velocity are experimental parameters to chose for which
the experimentalist must exercise judgement. Slight variations in
these parameters exist from laboratory to laboratory as do
variations in the method of column preparation. Columns prepared
by the methods described here have proven to be highly effective
for IGC experiments in the authors' laboratories.

Literature Cited

1. Lipson, J. E. G.; Guillet, J. E. Development in Polymer
    Characterization - 3; (Ed.) Dawkins, J. V. Applied Science
    Pub: 1982.
2. Jennings, W. Gas Chromatography With Glass Capillary Columns,
    2nd Ed.; Academic Press: New York, 1980.
3. Ward, T. C.; Sheehy, D. P.; McGrath, J. E.; Davidson, T. F.;
    Riffle, J. S. Macromolecules, 1981, 14(6), 1791-1797.
4. Al-Saigh, Z. Y.; Munk, P. Macromolecules, 1984, 17, 803.
5. Summers, W. R.; Tewari, Y. B.; Schreiber, H. P.
    Macromolecules, 1972, 5, 12.
6. Braun, J. M.; Guillet, J. E. Macromolecules, 1975, 8(6), 882.
7. Bolvari, A. Master's Thesis, Virginia Polytechnic Institute
    and State University, Virginia, 1988.
8. Jennings, W. Comparison of Fused Silica and Other Glass
    Columns in Gas Chromatography; Alfred Huthig Verlag, New York,
    1981.
9. Pawlisch, C. A.; Macris, A.; Laurence, R. L. Macromolecules,
    1987, 20, 1564.
10. Card, T. W.; Al-Saigh, Z. Y.; Munk, P. Macromolecules, 1985,
    18, 1030.
11. Guillet, J. E. Proceedings of the ACS Division of Polymeric
    Materials: Science and Engineering, 1988, 58, 645.

RECEIVED November 2, 1988

Chapter 3

Studies of Polymer Structure and Interactions by Automated Inverse Gas Chromatography

James E. Guillet, Marianne Romansky, Gareth J. Price[1], and Robertus van der Mark

Department of Chemistry, University of Toronto, Toronto, Ontario M5S 1A1, Canada

Inverse gas chromatography is a useful tool in the study of the structure of organic polymers and their interactions with a variety of permeating and adsorbing species. Its experimental simplicity allows the collection of data in a short time; however, its widespread use has been inhibited by difficulties in interpretation and the lack of commercially available instrumentation. Various aspects of IGC are reviewed, including the determination of glass and melting transition temperatures, degrees of crystallinity, solubility and interaction parameters and other thermodynamic quantities, surface areas, and adsorption isotherms of synthetic polymers. Several methods of automating IGC experiments are described, using conventional electronic and microcomputer control systems. The automated systems provide more reliable data, particularly in experiments requiring slope determinations, while also providing direct readout of the more important results.

Since its introduction in 1952 by James and Martin, the applications of gas liquid chromatography have grown enormously. This is due to the sensitivity, speed, accuracy, and simplicity of this technique for the separation, identification, and quantitation of volatile compounds.

The application of gas chromatography (GC) to the study of polymers has been hampered by their negligible volatility. A solution to this problem is the use of *inverse gas chromatography* (IGC, also called the *molecular probe technique*), which was developed by Smidsrød and Guillet (1) in 1969. The word "inverse" indicates that the component of interest is the stationary polymer phase, rather than the injected volatile substances.

Much of what is presently known about the structure and chemical interactions of macromolecules comes from physico-chemical studies in dilute solution, where the molecules are substantially

[1]Current address: School of Chemistry, University of Bath, Claverton Down, Bath, Avon BA2 7AY, England

isolated from each other. By contrast, in most practical applications, the polymer is concentrated and usually represents 90% or more of the bulk phase. Under these conditions, experimental techniques developed for dilute solution studies are frequently inapplicable. There is a large and increasingly important category of polymers that are insoluble in all known solvents and hence cannot be studied at all in solution.

Inverse gas chromatography (2) eliminates both of these difficulties. The polymer is studied in the solid phase under conditions approximating those used in processing and fabrication. Although a polymer may be insoluble in conventional solvents, virtually all small organic molecules have measurable solubilities in solid organic polymers, even when the latter are crosslinked or highly crystalline. Hence, the range of solute-solvent interactions that can be probed by IGC is virtually unlimited.

Considering the general availability of gas chromatographic equipment, the experimental simplicity, and the ease with which data can be collected, inverse gas chromatography is becoming the preferred method for the study of thermodynamic interactions of small molecules with polymers in the solid phase (3-9). However, the method is not limited to equilibrium measurements in the bulk phase. It can also be used to measure surface areas and adsorption isotherms (10-12), glass and other solid phase transitions in polymers (1,7,13-17), degrees of crystallinity (18-20) and diffusion constants for small molecules in polymeric materials (21-25). As the theory becomes more advanced, it is likely that other applications will develop, particularly in probing the structure of amorphous glasses.

Conventional Gas Chromatography (GC)

Gas chromatography is based on the distribution of a compound between two phases. In gas-solid chromatography (GSC) the phases are gas and solid. The injected compound is carried by the gas through a column filled with solid phase, and partitioning occurs via the sorption-desorption of the compound (probe) as it travels past the solid. Superimposed upon the forward velocity is radial motion of the probe molecules caused by random diffusion through the stationary phase. Separation of two or more components injected simultaneously occurs as a result of differing affinities for the stationary phase. In gas-liquid chromatography (GLC), the stationary phase is a liquid coated onto a solid support. The mathematical treatment is equivalent for GLC and GSC.

There are two mechanisms of gas-solid interaction to be considered: absorption of the solute in the bulk stationary phase, or adsorption on the surface of the stationary phase, or a combination of both. In conventional GC, the theory is based on bulk absorption. The net volume required to move the probe molecules through the column is V_N, the total volume of gas needed minus the "dead" (or simply spatial) volume in the column. This is determined by injecting an inert probe, such as methane or air, into the column (Figure 1). The parameter used in further calculations is V_g, the specific retention volume

$$V_g = (273.16/T)(V_N/w)(760/P_o) \qquad (1)$$

corrected to standard temperature and for the pressure drop across
the flowmeter, where $T(K)$ is the temperature of the column, P_o is
the column outlet pressure, and w is the weight (in grams) of polymer in the column.

Usually the retention volume is obtained using peak maxima to
define the retention times. In this treatment, since bulk absorption only is assumed, band broadening effects and the existence of a
non-linear sorption isotherm are not considered, as these usually
reflect some surface adsorption, resulting in skewed peaks.

Everett (26) developed the thermodynamic analysis for a binary
solution of components 1 (probe) and 2 (stationary phase) in the
presence of a gas (3), which is insoluble in the solution. Assuming
that the molar volume of the probe, V_1, does not vary greatly with
pressure, the gas phases are only slightly imperfect, the system is
in equilibrium, and the solute is infinitely dilute in both gas and
solid phases, then the infinite dilution mole fraction activity
coefficient of component 1 at temperature T and total pressure P can
be written as

$$\ln \gamma_1^\infty = \ln \left(\frac{n_L RT}{KV_L p_1^0} \right) - \frac{(B_{11}-V_1)p_1^0}{RT} + \frac{(2B_{13}-V_1^\infty)P}{RT} \qquad (2)$$

where n_L is the number of moles of component 2 occupying volume V_L
on the column, p_1^0 is the partial pressure of 1 in the vapor phase,
R is the gas constant, B_{11} is the second virial coefficient for the
probe, B_{13} is the mixed virial coefficient of the solute vapor and
carrier gas, V_1^∞ is the partial molar volume of 1 at infinite dilution, P is the total pressure, and K is the equilibrium partition
coefficient, defined as the ratio of concentration of solute in the
stationary phase, q, to that in the gas phase, c, that is, $K \equiv q/c$.

Literature values of experimental mixed virial coefficients are
scarce. At moderate carrier gas pressures (less than 2 atm), the
last term in Equation 2 can be ignored. Rewriting Equation 2 in
terms of the specific retention volumes gives

$$\ln \gamma_1^\infty = \ln \left(\frac{273.16\ R}{V_g p_1^0 M_2} \right) - \frac{(B_{11}-V_1)p_1^0}{RT} \qquad (3)$$

Other thermodynamic quantities can be calculated from the activity coefficient; for example, the excess free energy of mixing
at infinite probe dilution

$$\Delta G_m^e = RT \ln \gamma_1^\infty \qquad (4)$$

and the excess enthalpy of mixing

$$\frac{\partial ln\gamma_1^{\infty}}{\partial(1/T)} = \frac{\Delta H_m^e}{R} \tag{5}$$

Since the molecular weight of the polymer is often undetermined or has a wide distribution, the use of mole fraction activity coefficients has many inherent difficulties. Patterson et al. (27) proposed the use of weight fraction activity coefficients, which is now standard practice. It is recommended that all IGC data on polymers be reported this way.

The Flory-Huggins interaction parameter χ and the Hildebrand-Scatchard solubility parameter δ for the polymer may also be calculated using previously described procedures (3).

Inverse Gas Chromatography (IGC)

In IGC, the species of interest is the stationary phase, which usually consists of a polymer-coated support or finely ground polymer mixed with an inert support. This is in contrast to conventional analytical GC, where the stationary phase is of interest only as far as its ability to separate the injected compounds is concerned. Also, in IGC, usually only one pure compound at a time is injected.

Information from a molecular probe experiment is usually presented in the form of a retention diagram, that is, a plot of log V_g against $1/T(K)$. A sample curve for a semi-crystalline polymer is shown in Figure 2. The slope reversals are indicative of phase transitions. Such transitions had been noted (28) as early as 1965 for polyethylene (PE) and polypropylene (PP), but the first comprehensive study of polymer structure using IGC was done in 1968 by Smidsrod and Guillet (1) on poly(N-isopropyl acrylamide) (poly-(NIPAM)).

In the retention diagram shown in Figure 2, segment AB represents the polymer below its glass transition temperature, T_g (29). Retention of the probe in this region arises from condensation and adsorption of the probe onto the polymer surface, since the probe is unable to (significantly) diffuse into the bulk of the polymer. The slope of this straight segment is given by $(\Delta H_v - \Delta H_a)/2.3R$, where ΔH_v is the latent heat of vaporization of the probe and ΔH_a is the enthalpy of adsorption of the probe on the polymer surface. Segment B-C represents nonequilibrium absorption, and C-D represents equilibrium absorption of the probe into the amorphous polymer phase. Experimental curvature in C-D, due to an increase in the heat of vaporization of the probe with decreasing temperature, may be corrected using the extrapolation procedure of Braun and Guillet (30). Section D-F represents the melting process, and F-G represents solution of the probe in the molten polymer.

Knowledge of the amount of polymer in the column is necessary for accurate results. The amount of polymer that has been coated onto a support can be determined from Soxhlet extraction or by calcination of both coated and uncoated support, giving the weight percent of volatile material, and hence the weight of polymer. Laub et al. (31) and Braun et al. (32) examined the errors involved in using IGC for measuring thermodynamic parameters and found that the largest source of error was in the determination of the amount of

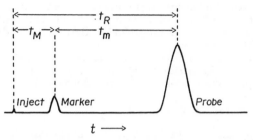

Figure 1. Typical gas chromatogram for a probe interacting with a polymer-coated stationary phase at temperature T. Retention volume $V_N = t_m$ x flow rate of carrier gas. (Reprinted with permission from ref. 3. Copyright 1982 Applied Science.)

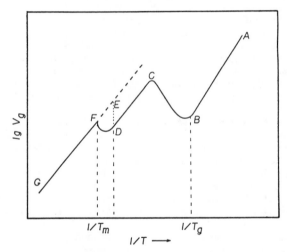

Figure 2. Retention diagram for a semi-crystalline polymer. (Reprinted with permission from ref. 3. Copyright 1982 Applied Science.)

polymer present; it was concluded that calcination (of silicon-free polymers) was preferred over extraction due to the presence of extractable inorganic materials in common supports.

Al-Saigh and Munk (33) have developed a soaking method for the coating of supports. Small amounts of solution are applied to a pile of support and allowed to evaporate. The wetting, evaporation, and stirring steps are repeated several times. Loss of polymer on the walls of the container vessel is avoided, and since the amount of polymer is precisely known, calcination or extraction is apparently not required.

A column can be packed with a polymer-support mixture or it may be coated on the inside with polymer to create a capillary column. Studies have been done using pure polymer in the column (34) and using capillary columns (35). Gray and Guillet found that V_g values for polystyrene (PS) were slightly higher for an open column (34) than for a packed column, possibly because of the higher specific surface area available in the open column. Chromosorb supports may increase V_g values, due to absorption of probes on some residual poly(dimethylsiloxane) (PDMS) caused by DMCS treatment (36). Lichtenthaler et al. (35) found that the capillary system was more sensitive to carrier gas flow rate. The V_g values differed from those obtained using the packed column by as much as 20% for poly(isobutylene) (PIB) and PDMS, and by more than 20% for poly(vinylacetate) (PVAc). In both cases, the difference decreased as the temperature increased. It was concluded that the basic disadvantage of the capillary method was the difficulty in calculating the amount of polymer present. Pawlisch et al. (24,37) have studied capillary column coating by a variety of methods, including scanning electron microscopy and destructive characterization. Based on their observations, they have improved the mathematical treatment for capillary columns to account for a nonuniform polymer film.

The development of modern microcomputers and associated instrumentation enables the automation of a number of IGC techniques. Automation is desirable because often 50 to 100 separate injections of very small volumes of probes are required over a period of time as the temperature of the GC is slowly increased, for example in the determination of transition temperatures or crystallinity. This paper will discuss the determinations of polymer crystallinity and the surface area of polymer-coated particles using automated instrumentation.

Determination of Polymer Crystallinity (18,19). The determination of polymer crystallinity from gas chromatographic retention data rests on the assumption that the probe molecules interact only with amorphous polymer; the crystalline regions are assumed to be impenetrable and do not contribute to the retention time of the probe. Therefore, the retention time is determined by the amount of amorphous material in the column. By extending the linear portion of the generalized retention curve to temperatures below the polymer melting temperature T_m, the hypothetical retention time t_a for a completely amorphous sample at any temperature may be obtained. Comparing this with the measured retention time t_m, at the same temperature gives the percentage crystallinity using the equation

$$\% \text{ crystallinity} = 100[1 - (t_m/t_a)]\qquad(6)$$

where t_a is determined by extrapolation to the temperature under study. An important feature of this method is that knowledge of the properties of 100% crystalline polymer is not required.

The basic information obtained from a typical chromatogram is shown in Figure 1. The determination of the retention time t_m for a typical probe, such as decane, is required as a function of temperature. Small amounts of the probe are injected onto the column, while the temperature is increased slowly using a temperature program.

In a typical experiment (19), data was obtained using the automatic system shown in Figure 3. The apparatus was based on a Varian Aerograph Model 1720 gas chromatograph equipped with a thermal conductivity detector. Helium was used as a carrier gas. At a preset cycle time, a mixture of nitrogen and decane vapor was introduced into the carrier gas stream using an electropneumatic injection system. The sample size was approximately constant for every injection in a series and was as small as practical. (All samples contained less than 1.5×10^{-6} mol of decane; no effect of sample size on retention time was apparent in this range.) The reported results were obtained with a 1 m $\times$ 0.25 inch o.d. copper tube packed with 0.16 g of high-density polyethylene (Tenite 3310, Tennessee Eastman Co.) coated on 60 to 80 mesh glass beads. The packing, containing 0.6% by weight of polymer, was sieved to 50 to 80 mesh before use.

The net gas chromatographic retention time for decane at a given temperature was measured by feeding the output from the thermal conductivity detector into an electronic peak detection system that measured the time between the peak maxima for nitrogen (noninteracting) and decane (interacting). The corresponding temperature was measured using an iron-constantan thermocouple attached to the outside of the gas chromatograph column. The net retention time and the temperature were recorded by a digital printer. The carrier gas flow rate, measured with a soap bubble flowmeter, was adjusted to given retention times between 10 and 500 s; retention times were reproducible to ±0.2 s at temperatures above the polymer melting points. Typical experimental data are shown in Figure 4. The polymer melting point corresponds to the cusp in the retention diagram.

At temperatures above the polymer melting point, a straight-line relationship was obtained. Using the automatic injection-detection system, the linearity was excellent. In a typical case for 34 data points between 140 and 200°C, the standard deviation in the slope was less than 0.2%.

The isothermal rate of crystallization can be followed by melting the polymer completely at a temperature above T_m, then reducing the column temperature to a point below T_m, and measuring t_m as a function of time. The maximum theoretical percent crystallinity at infinite time (at each temperature) is found by measuring t_e, the retention time when crystalline and amorphous regions have reached equilibrium during heating from room temperature. The ratio of percent crystallinity found to the maximum percent crystallinity yields the percent crystallization. Typical data for a high-density polyethylene sample are shown in Figure 5.

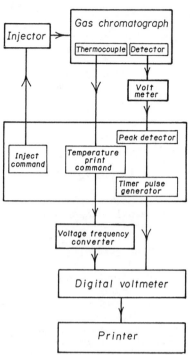

Figure 3. Automatic molecular probe apparatus. (Reprinted from ref. 19. Copyright 1971 American Chemical Society.)

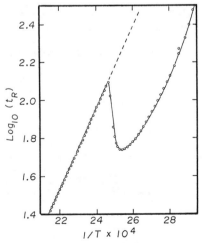

Figure 4. Retention diagram for decane on high-density polyethylene. T_m is at the cusp. (Reprinted from ref. 19. Copyright 1971 American Chemical Society.)

<u>Determination of Surface Areas</u> (<u>10</u>). At temperatures well below T_g, retention of probe molecules is primarily through adsorption at the polymer surface. The adsorption isotherm can be determined by the *elution technique*, where the probe is injected onto the column, and the shape of the isotherm is found from a single, asymmetrical peak (<u>10</u>). Using this method, it can be shown that for each gas-phase concentration c of a solute (mol/m^3) there is a corresponding retention volume V(c), m^3, according to

$$a = 1/m \int_0^C V(c)dc \qquad (7)$$

where a is the amount of solute (mol/g) absorbed on a mass m (g) of polymer.

In Figure 6 the elution peak shapes are shown for large injections of n-decane on poly(methyl methacrylate) (PMMA) at 25°C. The asymmetrical elution peak has a vertical and a diffuse (curved) side. V(c) can be determined from the diffuse region, and the isotherm relating a and c is thus determined via Equation 7 by procedures described previously (<u>10</u>).

A microcomputer-interfaced gas chromatograph was developed, illustrated in Figure 7, to simplify calculations and reduce the errors inherent in manual measurement of peak heights and areas from a chart recorder. The flame ionization detector (FID) analog output signal from a Carle AGC 211 gas chromatograph is amplified, then converted to a digital signal by an IBM data acquisition and control adapter (DACA) interfaced to an IBM-compatible personal computer. The GC oven temperature, monitored by a copper-constantan thermocouple thermometer, is also recorded by the microcomputer. Temperature stability is ±0.1°C. A computer program written in BASIC, combined with commercial data acquisition software (ISAAC Labsoft) controls data acquisition and calculations up to and including the surface area. The peak is divided into "slices" by the program, and cumulative partial and relative pressures are calculated along the diffuse side of the peak, as well as the amount of probe absorbed onto the polymer surface. Therefore, the absorption isotherm is calculated directly from the FID signal height and partial peak areas with corrections for dead volume, and a calibration for moles of probe per peak area unit. In a typical experiment, PMMA was coated onto glass beads using previously described procedures (<u>10</u>). Nitrogen carrier gas and methane marker were used. Decane was injected using a 0.5 or 1.0 μL syringe.

Typical isotherms from the automated system are shown in Figure 8. The data from one isotherm is fitted to the Brunauer-Emmett-Teller (BET) equation to obtain the surface area, and once the surface area is known, the surface partition coefficient K_s can be calculated using

$$V_N = K_s A \qquad (8)$$

where V_N is the net retention volume. Table I shows the values of surface area A and partition coefficients K_s, determined from the data of Figure 8, at various temperatures.

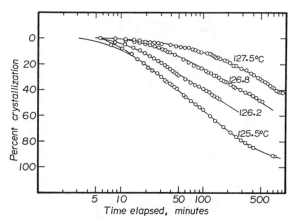

Figure 5. Percentage isothermal crystallization at the indicated temperatures as a function of time for high-density polyethylene. (Reprinted from ref. 19. Copyright 1971 American Chemical Society.)

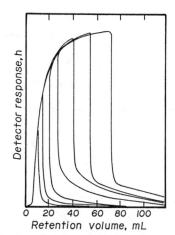

Figure 6. Peak shapes for large injections of n-decane on poly-(methyl methacrylate) beads at 25°C. Injection sizes 1.0, 0.7, 0.5, 0.3, 0.15, 0.06, 0.03 L (Reprinted from ref. 10. Copyright 1972 American Chemical Society.)

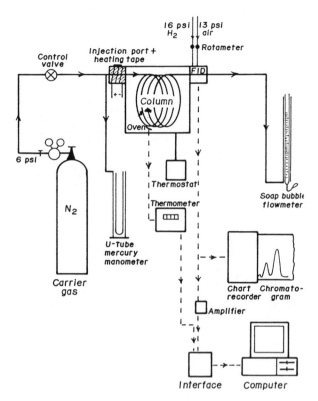

Figure 7.   The computer interfaced GC system.

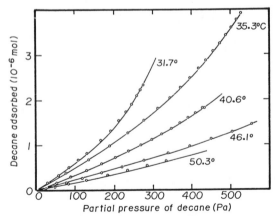

Figure 8.   Isotherms at the indicated temperatures for n-decane on
poly(methyl methacrylate)-coated glass beads

Table I*. Partition Coefficients for
Decane on PMMA

T (°C)	Surface area in column A, m^2	K_S 10^{-5} m^3/m^2
31.7	0.3021	7.7517
35.3	0.3119	5.3699
40.6	0.3054	4.1812
46.1	0.2909	3.0979
50.3	0.3114	2.3628

*Error in surface area is ±10%.

This method of determining partition coefficients is particularly useful because it does not require knowledge of the geometry or uniformity of the coating. Once the value of K_S for a particular probe and polymer system is known, the surface area of any powder can be estimated by packing a small column with a few hundred milligrams of dry powder, measuring the retention time for the probe at a temperature well below T_g, and applying Equation 8. This method of determining surface areas of polymer powders or particles has many advantages over classical procedures, such as nitrogen adsorption. One advantage is the speed with which the surface area may be found, and another is the capability of determining the area at the temperature of use of the glassy polymer, rather than at liquid nitrogen or helium temperatures.

Conclusions

From the examples given in this paper, it is evident that microprocessor control of the molecular probe experiment adds scope and precision to the measurements, as well as avoiding the tedium and operator error involved in repeatedly and reproducibly injecting small volumes of probes (for crystallinity determinations) and permitting quick calculations of adsorption isotherms and BET plots, in the case of surface work. It is hoped that instruments specifically designed for the purpose will be available commercially in the near future.

Literature Cited

1. Smidsrød, O.; Guillet, J. E. *Macromolecules* 1969, *2*, 272.
2. Guillet, J. E. *J. Macromol. Sci. Chem.* 1970, *A4*, 1669.
3. Lipson, J. E.; Guillet, J. E. in *Developments in Polymer Characterization* - *3*; Dawkins, J. V., Ed., Applied Science Publishers: Barking, 1982.
4. Price, G. J.; Guillet, J. E. *J. Macromol. Sci. Chem.*, 1986, *A23*, 1487.

5.   Price, G. J.; Guillet, J. E.; Purnell. J. H. *J. Chromatogr.*
     1986, *369*, 273.
6.   Price, G. J.; Guillet, J. E. *J. Solution Chem.* 1987, *16*, 605.
7.   Barrales-Rienda, J. M.; Gancedo, J. V. *Macromolecules* 1988, *21*,
     220.
8.   DiPaola-Baranyi, G.; Guillet, J. E. *Macromolecules* 1978, *11*,
     224.
9.   DiPaola-Baranyi, G.; Guillet, J. E. *Macromolecules* 1978, *11*,
     228.
10. Gray, D. G.; Guillet, J. E. *Macromolecules* 1972, *5*, 316.
11. Gozdz, A. S.; Weigmann, H.-D. *J. Appl. Polym. Sci.* 1984, *29*,
     3965.
12. Anhang, J.; Gray, D. G. *J. Appl. Polym. Sci.* 1982, *27*, 71.
13. Lavoie, A.; Guillet, J. E. *Macromolecules* 1969, *2*, 443.
14. Galin, M.; Guillet, J. E. *J. Polym. Sci., Polym. Lett. Ed.*
     1973, *11*, 233.
15. Braun, J.-M.; Lavoie, A.; Guillet, J. E. *Macromolecules* 1975,
     8, 311.
16. Sanetra, R.; Kolarz, B. N.; Wlochowicz, A. *Polymer* 1985, *26*,
     1181.
17   Tyagi, O. S.; Deshpande, D. D. *J. Applied Polym. Sci.* 1987, *34*,
     2377.
18. Guillet, J. E.; Stein, A. N. *Macromolecules* 1970, *3*, 102.
19. Gray, D. G.; Guillet, J. E. *Macromolecules* 1971, *4*, 129.
20. Sen, A. K.; Kumar, R. *J. Applied Polym. Sci.* 1988, *36*, 205.
21. Gray, D. G.; Guillet, J. E. *Macromolecules* 1973, *6*, 223.
22. Braun, J.-M.; Poos, S.; Guillet, J. E. *J. Polym. Sci., Polym.
     Lett. Ed.* 1976, *14*, 257.
23. Munk, P.; Card, T. W.; Hattam, P.; El-Hibri, M. J.; Al-Saigh,
     Z. Y. *Macromolecules* 1987, *20*, 1278.
24. Pawlisch, C. A.; Bric, J. R.; Laurence, R. L. *Macromolecules*
     1988, *21*, 1685.
25. Hu, D. S.; Han, C. D.; Stiel, L. I. *J. Polym. Sci.* 1987, *33*,
     551.
26. Everett, D. H. *Trans. Faraday Soc.* 1965, 1637.
27. Patterson, D.; Tewari, Y. B.; Schreiber, H. P.; Guillet, J. E.
     Macromolecules 1971, *4*, 356.
28. Alishoev, V. R.; Berezkin, V. G.; Mel'nikova, Y. V. *Russ. J.
     Phys. Chem.* 1965, *39*, 105.
29. Guillet, J. E. in *New Developments in Gas Chromatography*;
     Purnell, J. H., Ed., Wiley: New York, 1973.
30. Braun, J.-M.; Guillet, J. E. *Macromolecules* 1977, *10*, 101.
31. Laub, R. J.; Purnell, J. H.; Williams, P. S.; Harbison, M. W.
     P.; Martire, D. E. *J. Chromatogr.* 1978, *155*, 233.
32. Braun, J.-M.; Cutajar, M.; Guillet, J. E.; Schreiber, H. P.;
     Patterson, D. *Macromolecules* 1977, *10*, 864.
33. Al-Saigh, Z. Y.; Munk, P. *Macromolecules* 1984, *17*, 803.
34. Gray, D. G.; Guillet, J. E. *J. Polym. Sci., Polym. Lett. Ed.*
     1974, *12*, 231.
35. Lichtenthaler, R. N.; Liu, D. D.; Prausnitz, J. M.
     Macromolecules 1974, 7, 565.
36. Card, T. W.; Al-Saigh, Z. Y.; Munk, P. *Macromolecules* 1985, *18*,
     1030.
37. Pawlisch, C. A.; Macris, A.; Laurence, R. L. *Macromolecules*
     1987, *20*, 1564.

RECEIVED Februrary 22, 1988

Chapter 4

Computer Simulation of Elution Behavior of Probes in Inverse Gas Chromatography

Comparison with Experiment

Paul Hattam, Qiangguo Du[1], and Petr Munk

Department of Chemistry and Center for Polymer Research, University of Texas, Austin, TX 78712

In order to facilitate the analysis of the shape and position of elution curves in inverse gas chromatography, such curves were generated in a computer for many well-defined situations. The effects of diffusion in the gas phase, of slow diffusion in the polymer phase (compared to an instantaneous equilibration of the probe), and of surface adsorption (Langmuir type) were simulated. A set of evaluation guidelines was established and was applied to several model experiments.

The use of inverse gas chromatography (IGC) to study the properties of polymers has greatly increased in recent years (1,2). The shape and position of the elution peak contain information about all processes that occur in the column: diffusion of the probe in the gas and the polymer phases, partitioning between phases, and adsorption on the surface of the polymer and the support. Traditional IGC experiments aim at obtaining symmetrical peaks, which can be analyzed using the van Deemter (3) or moments method (4). However, the behavior of the polymer–probe system is also reflected in the asymmetry of the peak and its tail. A method that could be used to analyze a peak of any shape, allowing elucidation of all the processes on the column, would be of great use.

It is difficult to separate the effects of the various processes contributing to the shape and position of an experimental elution peak because, in most instances, it is not obvious which factors are at play in any particular experiment. Hence, it is useful to analyze various models of chromatographic processes theoretically and follow their effect on the elution peak. However, the differential equations describing these models

[1]Permanent address: Materials Science Institute of Fudan University, Shanghai, People's Republic of China

0097–6156/89/0391–0033$06.00/0

may be solved analytically only for the simplest models. It is possible to cast these differential equations in the form of difference equations and follow the development of the system by a computer. This paper reports the results of our computer simulations for some simple systems and compares the results with appropriate experimental data.

Traditional Analysis of Elution Curves

The commonly used method of analysis postulates that ideal elution curves are symmetrical and Gaussian. The time at the position of the peak maximum, t_R, is a measure of the distribution coefficient of the probe between the stationary and mobile phases; peak spreading is expressed by the height equivalent to one theoretical plate H, which may be written as $H = L/N$, N being the number of theoretical plates and L the column length. Furthermore we may write

$$N_P = (t_R/W_{\frac{1}{2}})^2 \, 8 \ln 2 \tag{1}$$

$W_{\frac{1}{2}}$ denotes the peak width at half height; and subscript P denotes a parameter obtained from peak dimensions. The extended (5) van Deemter equation (3) may be written in a general form as

$$H = A + 2\gamma D_g/u + (C_g + C_1)u \tag{2}$$

A is an eddy diffusion term to account for the various pathways in packed columns which lead to peak spreading; γ is the tortuosity factor which often has a value close to unity; u is the linear velocity of the carrier gas; terms C_g and C_1 account for radial diffusion in the gas phase and the liquid phase respectively. We have found experimentally on packed columns that the C_g term is negligible (6). The expression for the C_1 term may be written as

$$C_1 = Jd_f^2 k'/(1 + k')^2 D_1 \tag{3}$$

D_g and D_1 are the diffusion coefficients of the probe in the two phases. J is a numerical constant and is equal to $8/\pi^2$ according to van Deemter (3), and equal to 2/3, according to Giddings (7). d_f is the thickness of the polymer layer.

For the traditional model, the elution time at the peak maximum, t_R, is related to the capacity factor k' and partition coefficient K by

$$k' = K \, V_L/V_G \tag{4}$$

$$t_R = t_0(1 + k') = t_0(1 + K \, V_L/V_G) \tag{5}$$

Here t_0 is the elution time of the marker (ideal, non-retained probe); V_L and V_G are the volumes of the two phases within the column.

Another method of analysis of elution peaks is based on the statistical moments of the curve and was first proposed by McQuarrie (8). The first moment, F_M, is the center of gravity of the peak; it is equal to t_R for a hypothetical symmetrical peak. The second central statistical moment, S_M, is equal to the peak variance (σ^2); it is related to H as

$$H_M = L/N = L\,\sigma^2/t_R^2 = L\;S_M/F_M^2 \tag{6}$$

The subscript M refers to the method of moments. The method of moments is applicable to all chromatographic systems characterized by a linear partition isotherm (that is, for K = constant) irrespective of diffusion processes deforming the elution peak.

When the partition isotherm is nonlinear (typically when surface adsorption is involved), the elution peaks exhibit long tails. Tailing is also caused by slow diffusion of the probe in the polymer and by technical artifacts: mixing in the injection chamber, etc.

Computer Simulation

Diffusion of the probe in the gas and polymer phases, and adsorption on the support and on the polymer surface (both types of adsorption have nonlinear isotherms), simultaneously play an important role in IGC experiments and must be accounted for properly. An extensive computational program is planned to simulate the individual processes and to assess their influence on chromatographic behavior. In a recent paper, simulated behavior of three types of system was described (9). In the simplest case, only diffusion in the gas phase was operative. This case corresponds to elution of an ideal marker. Simultaneous effects of gaseous diffusion and partitioning of the probe between the phases were   simulated next, assuming an instantaneous equilibrium between the phases. This case corresponds to IGC using a low molecular weight stationary phase or a polymer well above its glass transition temperature. Simulation of the partition of the probe combined with its slow transport in the polymer phase and with gaseous diffusion was also performed.

It is convenient in dealing with the computer simulation to minimize the number of input variables governing the chromatographic processes. We were able to do this using just three characteristic numbers : Z_p for the partition of the probe at equilibrium, Z_g for the diffusion in the gas phase, and Z_f governing the diffusion of the probe in the polymer phase (it vanishes when the probe equilibrates instantly). These quantities are defined as

$$Z_p \equiv k' \equiv K\,V_L/V_G \tag{7}$$

$$Z_g \equiv D_g/uL \tag{8}$$

$$Z_f \equiv ud_f^2/D_1L \tag{9}$$

Should one wish, these variables can easily be converted back to their expanded form through the definitions given in Equations 7 - 9. Our measure of peak asymmetry is the ratio of half widths, $R_{\frac{1}{2}}$, (easily accesible experimentally); it is defined as a ratio of the

front half of $W_{\frac{1}{2}}$ to its back half. For symmetrical peaks $R_{\frac{1}{2}} = 1$.
Traditional analyses of elution peaks (Equations 2-6) in the
present notation are

$$t_R = F_M = (1 + Z_p)t_o \qquad (10)$$

$$H_P = H_M = 2Z_g + Z_f Z_p/(1 + Z_p)^2 \qquad (11)$$

We have simulated elution curves for many combinations of the
characteristic numbers. The details of the simulation procedure
were described elsewhere ($\underline{9}$). Here we present only the main
results of the simulations:

1. Elution peaks were always asymmetric, even for simple gaseous
diffusion and no interaction with the polymer. The $R_{\frac{1}{2}}$ values
were well correlated by the following expression:

$$R_{\frac{1}{2}} = 1 - (1.664(Z_g)^{\frac{1}{2}} + 1.225Z_g) \qquad (12)$$

This relation was also valid for interacting probes so long as the
equilibration was instantaneous.

2. The elution time t_R was always shorter than required by
Equation 10. For instantaneous equilibrium, the correlation
yielded

$$t_R/t_o = (1+Z_p)(1 - 2.77 Z_g) \qquad (13)$$

In this expression, t_o is the elution time of a hypothetical
marker with vanishing values of D_g and hence Z_g, which would
travel through the column as a Dirac delta function.

3. When the liquid diffusion was slow (large values of Z_f), the
probe eluted together with marker (that is, $t_R/t_o = 1$) and the
interaction with the polymer was manifested only by a long tail on
the elution peak.

4. When the ratio Z_f/Z_p was less than approximately 0.5, then a
pseudo-equilibrium was achieved, (the probe distributed itself
between the phases in a more or less equilibrium manner at least
near the end of the column), and t_R was given approximately as

$$t_R/t_o = (1+Z_p)(1-2.77Z_g) - 0.482Z_f(1+0.68Z_f/Z_p) \qquad (14)$$

5. Asymmetry of the elution peak $R_{\frac{1}{2}}$ and the value of H_p for slowly
diffusing probes depended on the ratio Z_f/Z_p. In the pseudo-
equilibrium case ($Z_f/Z_p < 0.5$), the asymmetry was moderate and
H_p was approximated by

$$H_p/L = 2Z_g + 0.7 Z_f Z_p/(1 + Z_p)^2$$
$$+ 0.965 [Z_f Z_p/(1 + Z_p)^2]^2 \qquad (15)$$

This relation is close to the van Deemter relation, Equation 10, so long as $Z_f Z_p/(1+Z_p)^2$ is not too large. When the ratio Z_f/Z_p is close to unity, R_L is small and H_p is very difficult to correlate with the basic parameters. Finally, when $Z_f/Z_p > 2$, the H_p value approaches the marker-like value of $2Z_g$ and the asymmetry decreases again.

6. The simulation results agreed fully with predictions of the moments method; both F_M and H_M were described by Equations 10 and 11. This was true even in the marker-like region, where the probes eluted essentially at $t_R = t_o$ and the retention was manifested only as a long low tail.

The second simulation project was aimed at describing experiments in which retention results from surface adsorption characterized by a Langmuir-type isotherm. In this simulation, the characteristic parameters are Z_g, Z_p, Z_s, and R_i. Z_s is the distribution coefficient between the surface and the carrier gas at vanishing surface coverage. R_i is defined as the ratio $R_i = M_{inj}/M_{tot}$ where M_{inj} is the mass of the probe injected and M_{tot} is the mass of the probe which would fully saturate the adsorbing surface. The surface simulation work shows that at infinite dilution

$$t_R/t_o = (1+Z_s+Z_p)(1 - 2.77Z_g) \tag{16}$$

$$F_M/t_o = (1+Z_s+Z_p) \tag{17}$$

With an increase in the amount injected (increasing R_i) both t_R/t_o and F_M/t_o decrease; t_R decreasing more rapidly than F_M. At injected amounts greater than the surface capacity of the column

$$t_R/t_o = (1 + Z_p)(1-2.77Z_g) \tag{18}$$

$$F_M/t_o = (1 + Z_p) \tag{19}$$

that is, the surface effect becomes negligible. Figure 1 illustrates results obtained from the surface simulation for the dependence of t_R/t_o and F_M/t_o on $Log_{10} M_{inj}$ ($M_{tot} = 1$) for several values of Z_s with $Z_p = 0$.

Materials and Methods

The experimental data presented in this paper represent typical data from chromatographic experiments that were performed during various IGC projects. The signal from the FID detector of the chromatograph was registered on an HP 3478A digital voltmeter and recorded by a microcomputer. The computer and voltmeter interfacing was performed by a GPIB interfacing board (National Instruments). Data acquisition in this manner allowed a reproducibility of approximately ±0.1 s in retention time. Typical columns were 150 cm long and 6.35mm O.D., and contained 60 to 80 mesh Chromosorb-W (acid washed and treated with dimethyl-dichloro-silane) either uncoated or coated with 7% (by weight) of polymer.

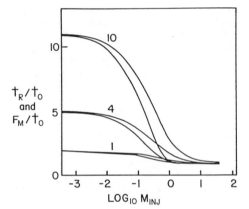

Figure 1. Dependence of t_R/t_0 and F_M/t_0 on $Log_{10} M_{inj}$ for the simulation of surface adsorption. $Z_g=0.02$; $Z_p=0$; $Z_s=1.0$, 4.0, 10.0.

Comparison of Simulation and Experiment

Demonstration of Instant Equilibration ($Z_f = 0$). Data were gathered for several n-alkanes at various gas flow rates and column temperature of 100°C using a column coated with poly-isobutylene (PIB). Under these conditions PIB is far above its glass transition temperature (Tg) and equilibration of probe and polymer is expected to be instantaneous.

For instantaneous equilibration, the simulation predicts that F_M is as predicted by theory, Equation 10, and t_R is given by equation 13. Thus t_R should be slightly less than F_M due to the gaseous diffusion coefficient. Table I shows the experimental values of t_R and F_M at several flow rates.

Table I. Flow rate dependence of t_R and F_M determined at 100°C for several n-alkanes on a poly-isobutylene column

Probe	8 mL/min		16 mL/min		24 mL/min	
	t_R	F_M	t_R	F_M	t_R	F_M
C1	176.736	178.289	91.400	91.929	64.700	64.990
C5	215.807	219.211	110.932	111.689	78.680	79.277
C6	258.432	259.945	133.056	133.880	94.213	94.838
C7	346.817	348.676	178.822	179.847	126.335	127.114
C8	527.889	529.752	271.797	272.727	192.487	193.384

The difference between t_R and F_M is less than 1%, which is reasonable for the expected magnitude of Z_g. Two points indicate the system is in equilibrium. First, there is close agreement between t_R and F_M. Second, $R_{\frac{1}{2}}$, the width ratios for the probes measured under identical conditions for a coated and an uncoated column are essentially the same, see Table II.

Table II. The width ratio, $R_{\frac{1}{2}}$, obtained for several n-alkanes on an uncoated column and on a PIB column at 40°C and 100°C at a flow rate of 16 mL/min

Probe	$R_{\frac{1}{2}}$ Chrom. W 40°C	$R_{\frac{1}{2}}$ Chrom. W 100°C	$R_{\frac{1}{2}}$ PIB 40°C	$R_{\frac{1}{2}}$ PIB 100°C
C1	0.951	0.942	0.949	0.946
C2	0.921	0.961	0.739	0.951
C3	0.918	0.963	0.796	0.923
C4	0.895	0.936	0.845	0.942
C5	0.817	0.942	0.888	0.946

By reducing the t_R of the probe by t_R of the marker (t'_R), and reducing the F_M of the probe by F_M of the marker (F'_M), $1+Z_p$ for the probes is obtained. (Reduction in this manner accounts for part of the small error in t_R because of the effect of Z_g.) The reduced values ($1+Z_p$) are shown in Table III. There is excellent agreement between the values calculated from either t_R or F_M. In addition, the values are independent of flow rate, as is expected for instantaneous equilibration.

Table III.   Reduced values of t_R and F_M determined for several n-alkanes at 100°C on a PIB column

Probe	8 mL/min		16 mL/min		24 mL/min	
	t_R/t'_M	F_M/F'_M	t_R/t'_R	F_M/F'_M	t_R/t'_R	F_M/F_M
C5	1.220	1.229	1.215	1.215	1.216	1.219
C6	1.461	1.457	1.458	1.456	1.455	1.458
C7	1.961	1.954	1.959	1.956	1.954	1.955
C8	2.985	2.970	2.978	2.966	2.974	2.973

Demonstration of Non-equilibrium ($Z_f/Z_p < 0.5$).   Data were gathered on the same PIB coated column as above at a temperature of 40°C. Under these conditions, diffusion of the probes into the polymer is not expected to be instantaneous. The simulation under these conditions predicts that $F_M = (1 + Z_p)$ and t_R will be reduced by the effect of non-zero Z_f. The magnitude of the reduction is given by the second term of the right side of Equation 15. Since Z_f is dependent on flow rate, it is possible to estimate Z_f from the dependence of the peak width on the flow rate and hence, determine the size of the correction and compare it with experimental results. This comparison should be made bearing in mind that theoretically the simulation is applicable for capillary columns and not packed columns.

First, the true value of Z_p is determined by extrapolation of V_n/V_o to zero flow rate, where V_n is the net retention volume of the probe and V_o is the void volume of the column. These values are shown in Table IV. Second, u, the linear velocity of the carrier gas, is introduced into Equation 11 to give

$$u/N_P = 2Z_g u + (0.7Z_f Z_p/(1 + Z_p)^2)u \qquad (20)$$

Substitution for Z_g and Z_f leads to

$$u/N_P = 2D_g/L + (0.7d_f^2 Z_p/D_l L(1 + Z_p)^2)u^2 \qquad (21)$$

Thus a plot of u/N_P versus u^2 yields

$$\text{intercept} = 2D_g/L \qquad (22)$$

$$\text{slope} = 0.7\ d_f^2 Z_p/(1+Z_p)^2\ D_l L \qquad (23)$$

Rearrangement gives

$$Z_f \equiv d_f^2 u/D_l L = \text{slope } (1+Z_p)^2 u/0.7\ Z_p \qquad (24)$$

The correction predicted by Equation 15 is estimated and Z_p is adjusted. Z_p at zero flow rate, the experimental values and the corrected values for three flow rates, are given in Table IV.

Table IV. Z_p values of n-alkanes on a PIB column at 40°C; values
at zero flow rate, at several flow rates, and at several
flow rates corrected according to the simulation

Flow (mL/min)	Z_p 0	Z_p from t_R/t'_R 8	16	24	Z_p corrected 8	16	24
Probe							
C5	0.872	0.832	0.800	0.777	0.842	0.822	0.810
C6	2.431	2.369	2.328	2.295	2.384	2.357	2.338
C7	6.700	6.600	6.546	6.511	6.681	6.678	6.692
C8	18.34	18.15	18.09	18.07	18.18	18.13	18.13

The predicted correction term is approximately three-fold less
than would be required to adjust the experimental values to their
value at zero flow rate. This difference is likely due to the
range of polymer thickness in packed columns rather than the
homogeneous coverage used in the simulation or for capillary
columns. However, the first moment yields Z_p values that are in
excellent agreement with Z_p values calculated at zero flow rate;
Table V.

Table V. The Z_p values of several n-alkanes determined on a PIB
column at 40°C; values at zero flow rate and values
determined from the first moments

Flow(mL/min)	Z_p 0	Z_p from F_M/t'_R 8	16	24
Probe				
C5	0.872	0.887	0.894	0.896
C6	2.431	2.430	2.444	2.435
C7	6.700	6.681	6.674	6.692
C8	18.34	18.31	18.30	18.27

This indicates that under these conditions, Z_p can be obtained
from t_R only by extrapolating to zero flow rate, whereas F_M may be
used regardless of the flow rate.

Determination of the Statistical Moments. The simulation
confirmed that the method of moments offers a straight forward
route to the data of interest for chromatographic experiments when
isotherms are linear. However, in the past the experimental
evaluation of the moments was imprecise. The statistical moments
are extremely sensitive to tailing. Using the enhanced data
acquisition techniques (signal to noise ratios of approximately
5×10^4) the method of moments was re-examined. The presence of a
long low tail on many of the elution peaks was observed on coated
as well as uncoated columns. It was also noted that a long low
tail was observed whenever the probe was injected in liquid form.
When the injected amounts were the same, vapor injections greatly
reduced tailing compared with liquid injections. The tailing is
attributed to retention of the probe by the polymeric septum of
the injection port. When the needle is inserted through the
septum, the liquid at the tip of the needle is transferred to the

septum. The probe slowly eluting from the septum causes excessive
tailing. Figure 2 illustrates the difference between the moments
obtained from liquid injections and vapor injections. The use of
vapor injections almost completely suppresses the effect of
retention by the septum. The moments are extremely sensitive to
tailing; the higher the moment, the greater the sensitivity. With
small injections of vapors we have found that the first moment can
be measured with confidence. Ideally a septum free system, should
be used for the introduction of probes into the column if the
higher moments are to be utilized. It is possible that headspace
sampling gas chromatography could be used advantageously for this
purpose ($\underline{10}$).

Peak Asymmetry. The asymmetry of the elution curve reflects the
various processes occurring in the column. To follow this
asymmetry width ratio, $R_{\frac{1}{2}}$, defined earlier, was used. Although no
quantitative relationship that can be applied with confidence
experimentally was found via the simulation, $R_{\frac{1}{2}}$ has proved to be a
useful quantity. In the case of the simulation of surface
adsorption at infinite dilution, $R_{\frac{1}{2}}$ is close to unity. As
injection size increases, $R_{\frac{1}{2}}$ decreases until $R_i = 1$. $R_{\frac{1}{2}}$ then
increases again as the probe begins to elute in the marker-like
region. It is hoped that in the future research will determine
whether the dependence of $R_{\frac{1}{2}}$ on probe concentration can be used in
determining support surface area. In the simulation of bulk
diffusion at instant equilibration $R_{\frac{1}{2}}$ is close to unity. It
decreases through the non-equilibrium region and again increases
as marker-like behavior is observed. Though the processes that
affect $R_{\frac{1}{2}}$ in a particular experiment cannot be determined, $R_{\frac{1}{2}}$ can
be used as a guideline. For example, Table II shows $R_{\frac{1}{2}}$ determined
for several n-alkanes on the PIB column and on an uncoated column
at 40°C and 100°C and a flow rate of 16 mL/min. At 100°C, $R_{\frac{1}{2}}$ for
probes on both uncoated and PIB columns are comparable and
relatively large. This indicates that there is no anomalous
behavior in the system. However the data for probes on the PIB
column at 40°C show a considerably lower $R_{\frac{1}{2}}$, indicating that the
system is not exhibiting instantaneous equilibration.

Acetone on Uncoated Support. Experiments in this section
were performed on a column containing uncoated support at 40°C and
100°C and a flow rate of 16 mL/min. After treatment of the
support with dimethyl-dichloro-silane, the resultant, so-called
inert support, still contained a small number of active polar
sites. These sites lead to adsorption of polar probes and thus
the support contributes to the observed retention. One of the
goals of this investigation was to facilitate the correction of
retention data for the contribution of the support. The
dependence of the acetone retention on the quantity injected
(determined from peak area) was investigated. In the absence of
polymer, any change in the retention with change of probe
concentration was expected to be the result of surface adsorption.
Also, by changing the temperature Z_s was effectively altered,
(increasing temperature leading to a decrease of Z_s), since the
surface capacity of the column remained the same. Figure 1

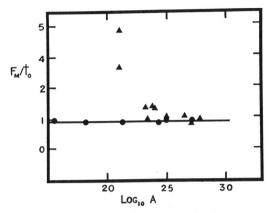

Figure 2. Dependence of the first moment on the amount of probe injected. (▲) probe injected as a liquid; (●) probe injected as a vapor.

illustrates the dependence of t_R/t_o and F_M/t_o on $\log_{10} M_{inj}$ for the simulation at several Z_s values and with $Z_p = 0$. In Figure 3 we present the same dependence for the experimental data of acetone (in this instance the dependence is on $\log_{10} A$, peak area). While the simulation covers an extensive range of concentrations this may not be possible experimentally. Small injections are limited by the low detector signal (broad peaks magnifying this effect), while large injections are limited by non-linear detector response. However, even with the experimental range of concentrations covered, a comparison of the curves in Figures 1 and 3 is useful. It indicates that the results are in the region of moderately sized injections (for the number of active sites on the column). At 100 °C Z_s is small, aproximately 1 or 2, whereas at 40 °C Z_s is greater by an order of magnitude. Although the precise value of Z_s cannot be determined as the injection size is far from the limit of infinite dilution, the value of F_M/t_o approaches unity with increasing amount of probe. This trend indicates that the retention is due to the surface of the support. Overall, the surface simulation and the experimental results compare favorably. Current work is focused on acquiring more data, both from experiment and from simulation. At this time it seems likely that in the future it will be possible to determine the capacity of the packing material for various probes. This will permit correction of probe retention data for the effect of the active surface sites on the support.

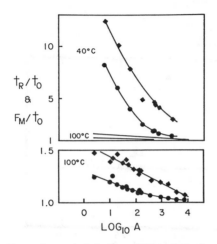

Figure 3.   Dependence of t_R/t_O (●) and F_M/t_O (♦) on $\text{Log}_{10} A$, peak area, for acetone on an uncoated column at 40°C and 100°C. Data at 100°C plotted twice for comparison.

Conclusions

The following conclusions were drawn from this research.

1. While the simulations do not predict exactly the results of experiment, they are extremely useful in predicting behavior trends.

2. The first moment can be used in the determination of characteristic numbers provided careful data acquisition and experimental procedure are followed. Use of higher moments should be handled with great caution.

3. Comparison of the elution time at peak maximum and the first moment is extremely informative as to what processes are affecting the retention of the probe.

4. By following the dependence of elution parameters on the amount of probe injected, it is possible to distinguish between surface adsorption and bulk adsorption of the probe.

Acknowledgment

The authors are grateful for the financial support of the National Aeronautics and Space Administration, (Grant No. NAG9-189) and the National Science Foundation, (Grant No. DMR-8414575).

Literature Cited.

1. Laub, J. R.; Pecsok, R. L. Physicochemical Applications of Gas Chromatography; Wiley: New York, 1978.
2. Aspler, J. S. In Pyrolysis and GC in Polymer Analysis; Chromatographic Science Series, Vol. 29: Liebman, S. A.; E. J. Levy Eds.; Dekker: New York, 1985. Chapter IX.
3. van Deemter, J. J.; Zuderweg, F. J.; Klinkenberg, A. Chem. Eng. Sci. 1956, 5, 271.
4. Vidal-Madjar, C.; Guiochon, G. J. Chromatogr. 1977, 61, 142.
5. Golay, M. J. E. Gas Chromatography;Coates, V. J.; Noebles, H. J.; Fagerson, I. S. Eds.; Academic Press, New York 1958.
6. Munk, P; Card, T. W.; Hattam, P.; El-Hibri, M. J.; Al-Saigh, Z.Y. Macromolecules 1987, 20, 1278.
7. Giddings, J. C. J. Chromatogr. 1961, 5, 49.
8. McQuarrie, M. J. J. Chem. Phys. 1963 38 ,437.
9. Hattam, P.; Munk, P. Macromolecules 1988, 21,2083.
10. McNally, M. E.; Grob, R. L. Amer. Lab. 1985, 17, 106.

RECEIVED November 2, 1988

SORPTION AND DIFFUSION IN POLYMERS

Chapter 5

Calculation of Solubility Parameters by Inverse Gas Chromatography

Gareth J. Price[1]

Department of Chemistry, The City University, Northampton Square, London EC1V 0HB, England

Inverse Gas Chromatography (IGC) has been used to measure solubility parameters for three polymers at 25°C using the method of Guillet and DiPaola-Baranyi. The linear relationship noted with other polymers was found and the results add further credance to the method. Solubility parameters have also been calculated for six small molecule involatile compounds of the type use as plasticizers. The original method did not yield values in good agreement with literature results but estimation of the different contributions to the solution interactions allowed calculation of more meaningful values.

The study of polymer solutions has been an active research field since the mid 1960s. There are a number of methods available for the measurement of thermodynamic parameters such as activity coefficients and interaction parameters [1,2]. These techniques, which include membrane osmometry and vapour sorption, involve difficult and time consuming experiments and are usually confined to relatively dilute solutions although vapor sorption using electronic vacuum microbalances, has been used at high polymer concentrations [3]. Inverse Gas Chromatography (IGC) is a method that overcomes these limitations [4,5] and is particularly applicable to concentrated solutions, which are of considerable industrial interest for surface coatings, solvent removal etc. Since the early work of Smidsrod and Guillet [6] in 1969 numerous systems have been studied by this method, and good agreement with the more traditional, static equilibrium measurements of activity coefficients, interaction parameters, enthalpies of mixing and solution and contact energy parameters has been demonstrated [7,8]. Another useful facet of the method is that it may be extended to the study of mixtures of two or more polymers to obtain information on polymer-polymer interactions [9,10] and also to the study of mixtures of polymers with lower molecular weight compounds such as plasticizers [11].

Paralleling this experimental work has been considerable activity in the theoretical treatment of polymer solutions. The original work of Flory and Huggins is often used for the calculation

[1]Current address: School of Chemistry, University of Bath, Claverton Down, Bath, Avon BA2 7AY, England

0097–6156/89/0391–0048$06.00/0
© 1989 American Chemical Society

of the interaction parameter, χ, now regarded as a residual free energy function. Other developments in the interpretation of χ have been the Corresponding States theory of Prigogine and Patterson [12, 13], Flory's Equation of State treatment [14], the Lattice Fluid method of Sanchez and Lacombe [15] and, more recently, the Scaling Concepts of de Gennes [16]. Although these treatments have allowed a more rigorous interpretation of the various parameters, they depend on a number of empirical parameters which cannot be readily predicted; therefore, they are of limited use in cases where little or no experimental data is available. Recourse must often be made to less rigorous, but more easily applied methods. Amongst the most often used is that using the Hildebrand solubility parameter, δ, [17,18]. Although of very limited theoetical significance, the concept remains useful for many practical applications such as solvent selection and the prediction of phase equilibrium.

The solubility parameter, δ, derived from the cohesive energy density, δ^2, used as a measure of intermolecular forces may be defined [19] as

$$\delta^2 = \frac{\Delta U^{vap}}{V^\circ} = \frac{\Delta H^{vap} - RT}{V^\circ} \tag{1}$$

where ΔU^{vap} and ΔH^{vap} are the molar internal energy and enthalpy of vaporization and V° is the molar volume of the liquid. This allows estimation of δ for small molecule liquids. However, this definition is not applicable for polymers and other involatile compounds and methods such as swelling equilibria or group contribution methods must be used. DiPaola-Baranyi and Guillet [20] developed a chromatographic method for the calculation of the solubility parameter of polymeric stationary phases, δ_2, from measurements of interaction parameters.

A frequent use of solubility parameters is the prediction of compatibility of blends of polymers with additives such as plasticizers used to modify the polymer properties. Plasticizers are generally involatile organic molecules such as dialkyl phthalates. Thus it was of interest to determine the usefulness of ICC method for estimating the solubility parameters of these compounds.

Polymer Solubility Parameters

The interpretation of the Flory-Huggins interaction parameter as a residual free energy function [14] rather than the original enthalpy parameter allows separation into enthalpic and entropic contributions

$$\chi = \chi_H + \chi_S \tag{2}$$

The method of DiPaola-Baranyi and Guillet is an extension of the work of Bristow and Watson [21] who calculated solubility parameters for a series of network polymers from swelling equilibria. The basis is that the solubility parameters of solvent, δ_1, and polymer, δ_2, are introduced in the form of Regular Solution theory [19] to account for enthalpic effects.

$$\chi^\infty = (V^\circ/RT)(\delta_1 - \delta_2)^2 + \chi_S^\infty \tag{3}$$

where the superscript ∞ indicates that IGC data are measured at infinite dilution of solvent in the polymer. Expansion of the term in
parentheses and rearrangement yields

$$\left(\frac{\delta_1^2}{RT} - \frac{\chi^\infty}{V^o} \right) = \left(\frac{2\,\delta_2}{RT} \right)\delta_1 - \left(\frac{\delta_2^2}{RT} + \frac{\chi_s^\infty}{V^o} \right)$$ (4)

A plot of the function on the left hand side of Equation 3 versus δ_1
should give a linear graph with the δ_2 value being calculated from
the slope.

The method, originally applied to polystyrene and polybutyl
methacrylate [20], has been applied to numerous polymers. Some of the
results are shown in Table I. In the majority of cases the δ_2 values
estimated at 25 °C agree very well with those calculated by traditional methods. A great advantage of the IGC method over other techniques is that measurements can be made considerably above room temperature. However, equilibrium thermodynamic measurements can only be
made at temperatures approximately 50 °C above the glass transition
temperature, Tg, of the polymer [22]. Thus, in the above work, interaction parameters were measured over a range of temperatures around
those indicated in Table I and extrapolated to 25 °C assuming a linear relation employing an equation of the form:

$$\chi = a + b/T$$ (5)

where a and b are constants for each polymer-probe system.

TABLE I. Solubility Parameter Values $(MPa)^{\frac{1}{2}}$ from IGC

POLYMER	Temperature	$\delta(t)^a$	$\delta(25)^b$	$\delta(Lit.)^c$	Ref.
polystyrene	193	15.5	19.8	17.4 – 21.5	20
polystyrene	140	15.3	18.6	17.4 – 21.5	23
polymethylacrylate	100	17.4	20.3	20.0 – 21.3	20
polyvinyl alcohol	125	18.0	21.7	19.0 – 22.7	23
polyvinyl chloride	125	16.2	18.8	19.2 – 22.1	24
polyethylene oxide	70–140	–	20.9		25
polyvinyl acetate	135	17.4	20.7	18.0 – 22.5	26
polypropylene	63–83		15.8	15.8 – 18.0	27
polyisoprene	63–83		16.4	16.2 – 17.0	27
polybutylmethacrylate	140	14.7	17.4	17.8	23
polychloroprene	75	18.0		16.8 – 19.2	28
poly-1,4-butadiene	75	16.2		14.7 – 17.6	28
poly(butadiene-acrylonitrile)	75	20.4		20.0 – 20.9	28
poly(ethylene-vinyl acetate)	75	17.0		17.2 – 19.8	28

a. Solubility parameter at temperature of measurement (°C).
b. Solubility parameter extrapolated to 25 °C.
c. Solubility parameters taken from literature values in the
   reference given in the final column.

This extrapolation procedure is necessary to compare the results with those from other methods which are usually measured at or near room temperature [31]. To test the validity of this procedure, δ_2 values have been calculated for three polymers using results measured directly at 25 °C rather than the higher temperatures employed previously. The polymers were polydimethylsiloxane (Tg = -150 °C), polyisobutylene (Tg = -200 to 210 °C) and ethylene-propylene rubber (Tg = -150 to 180 °C). The retention volumes and interaction parameters have been given elsewhere [30]. The plots derived from Equation 3 are shown in Figure 1 with the calculated δ_2 values compared with literature results listed in Table II. Again, the results are in excellent agreement with the predicted linear relation and also with the literature values adding further validity to the IGC estimation of δ_2.

TABLE II. Solubility Parameters of Polymers (MPa)$^{\frac{1}{2}}$ at 25 °C

	Current Work	Literature [31]
Poly(dimethylsiloxane)	15.10	14.9 - 15.5
Polyisobutylene	16.16	15.8 - 16.6
Ethylene-Propylene Rubber	16.63	16.2 - 17.2

One notable feature of this work, and of all the polymers so far investigated, is the excellent correlation of the results with the linear relation predicted by Equation 3 despite the obvious approximations underlying the method and the various types of polymer studied. This has inevitably prompted further speculation and comment. A particularly intriguing aspect is that the same results have been obtained irrespective of the polarity of the polymer and solvents so that the nature of the intermolecular forces in the solution appear to have little effect [32].

In particular, Lipson and Guillet [28] have commented at length on the significance of the X_S parameter and have attempted to correlate its value with properties of particular systems, but no systematic pattern emerged for its contribution to X . Recently, Price, Guillet and Purnell [30] suggested that the X value as measured by IGC, reflected contributions to the Helmholtz free energy of the mixing process, based on changes of internal energy rather than enthalpy. The fact that the solubility parameter is also an internal energy function suggests that X_H accounts for these differences and X_S, which is left to mop-up all other contributions to the overall free energy change, contains entropic and pressure-volume effects.

Small Molecule Solubility Parameters

It was of interest to determine whether the same considerations outlined above would be applicable to systems involving small molecules such as those used as polymer additives. Much of the early work, showing the utility of IGC for the measurement of thermodynamic parameters, was performed on this type of compound. Consequently, there is a large number of results in the literature analysed using the Flory-Huggins theory in a manner analogous to polymer systems [33]. The treatment outlined above has been applied to six representative compounds of this type. Those chosen were two long chain, non-polar alkanes, n-hexadecane ($C_{16}H_{34}$), and squalane ($C_{30}H_{62}$); two compounds containing polar groupings, N-methyl pyrrolidone and dibutyl-2-ethyl hexamide; and two alkyl phthalates of the type used as plasticizers, dinonyl phthalate (that is, the 3,5,5-trimethyl hexyl isomer) and di-n-octyl phthalate. These are abbreviated as HEX, SQ, NMP, DBEH, DNP and DOP respectively. In most cases, the results were presented in the literature as χ parameters although in the case of NMP, activity coefficients were reported and χ values were calculated using literature data following the usual procedure as outlined in Reference 33.

The plots suggested by Equation 3 for these systems are shown in Figures 2, 3 and 4 and these show that the excellent correlation found with polymers is not obeyed with these compounds. The results arising from alkane probes show linearity, even with polar molecules such as NMP, but there are marked deviations with aromatic or polar probes. Table III shows the solubility parameters calculated using solely the linear portions of the graphs obtained with the alkane probes, and also using the results from all of the probes. The results do not show the consistency displayed by polymers, and the estimates of δ_2 differ greatly from literature values. There is no apparent pattern as to whether better results are found from the alkane solvents or from all of the results.

TABLE III. Solubility Parameters of Involatile Compounds

	TEMP. °C	SOURCE REF.	SOLUBILITY PARAMETERS, δ (MPa)$^{\frac{1}{2}}$		
			All[a]	Alkane[b]	Literature[c]
HEX	30	34	14.1	15.8	16.0
SQ	30	35	14.3	16.2	15.8
NMP	30	36	22.3	16.2	23.1
DBE	30	34	21.0	16.2	17.4
DNP	30	35	18.8	16.2	17.2
DOP	75	11	18.0	16.7	16.2

a. Calculated from all points on plots in Figures 2 to 4.
b. Calculated from linear portions of plots in Figures 2 to 4.
c. NMP and DOP from Reference 18; DNP from Reference 37; SQ and HEX estimated from properties of similar compounds; DBE estimated from Small's group contribution method in Reference 31.

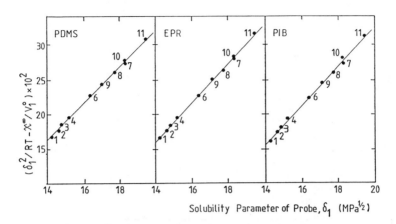

Figure 1. Calculation of polymer solubility parameters at 25°C for poly(dimethyl siloxane), PDMS, ethylenepropylene rubber, EPR, and polyisobutylene, PIB.

Probes: 1. n-pentane; 2. n-hexane; 3. n-heptane; 4. n-octane; 5. methyl cyclohexane; 6. cyclohexane; 7. benzene; 8. toluene; 9. carbon tetrachloride; 10. chloroform; 11. dichloromethane.

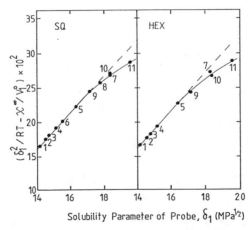

Figure 2. Calculation of solubility parameters for squalane (SQ) and n-hexadecane (HEX) at 30°C.   (Probes as in Figure 1.)

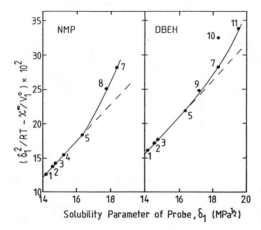

Figure 3. Calculation of solubility parameters for N-methyl pyrrolidone (NMP) and N,N-dibutyl ethyl hexamide (DBEH) at 30°C. (Probes as in Figure 1.)

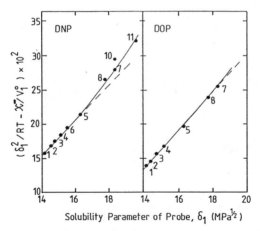

Figure 4. Calculation of solubility parameters for dinonyl phthalate (DNP) at 30°C and dioctyl phthalate (DOP) at 75°C. (Probes as in Figure 1.)

Discussion

Even with the long chain alkane molecules HEX and SQ, there is pronounced curvature in the plots, although these systems might be expected to be free from any polar interactions. The curvature in the
alkane systems is downward, leading to low estimates of δ_2 while the
curvature in the more polar compounds was upward, leading to overestimates. It may be of interest to speculate why these compounds
show curvature while structurally similar polymers, such as ethylene-
propylene rubber and polypropylene, give linear plots.
   One possible explanation lies in the nature of the intermolecular forces involved. Amongst the most popular extensions to the
basic solubility parameter theory is that due to Hansen [38] which
considers contributions from three types of intermolecular forces:

$$\delta^2 = \delta_d^2 + \delta_p^2 + \delta_h^2 \qquad (6)$$

where δ_d arises from dispersion forces, δ_p from polar forces and δ_h
from hydrogen bonding. In the current work, the latter contribution
may be neglected, so that

$$\delta^2 = \delta_d^2 + \delta_p^2 \qquad (7)$$

Thus, certain polymer-solvent combinations might be expected to give
rise to curved plots. However, as noted above, this has not thusfar
been found and it appears that the 'Three Dimensional' approach is
not useful for polymer δ_2 values calculated by this method.
   In an attempt to separate the various contributions to the small
molecule solubility parameters the slopes of the graphs in Figures 2
to 4 were calculated separately using aromatic and aliphatic probes.
The latter of these was assumed to account for δ_d while the difference between them was ascribed to δ_p. An overall value of δ was then
calculated from Equation 7. The procedure is illustrated in Figure 5
using NMP as an example. The results for all the liquids are
summarized in Table IV.

TABLE IV. Two Dimensional Treatment for Solubility Parameters

| | SOLUBILITY PARAMETERS, $\delta(MPa)^{\frac{1}{2}}$ | | | |
	δ_d	δ_p	δ	Literature
HEX	15.8	−4.1	16.0	16.0
SQ	16.2	−5.7	17.0	15.8
NMP	16.2	15.3	22.3	23.1
DBE	16.2	6.8	17.6	17.4
DNP	16.2	3.5	16.6	17.2
DOP	16.2	2.5	16.8	16.2

   As may be seen, the prediction of δ for the polar compounds is
considerably improved by this procedure, implying that there may well
be some merit in the separation of the contributions to δ.

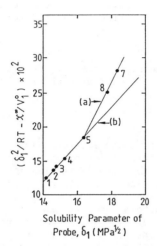

Figure 5. Calculation of the contributions to the solubility parameter for N-methyl pyrrolidone.

(a). slope due to δ_d and δ_p; (b). slope due to δ_d only.
(Probes as in Figure 1.)

However, there remain a number of unanswered questions raised by this work. For instance, the "negative δ_p" values for SQ and HEX cannot be explained by straightforward solubility parameter theory. Similarly, the grouping of alkane probes into one 'family' and the aromatic and chloroalkane probes into another may not be justified. Interpretation in terms of acid-base behaviour could be attempted but it is difficult to envisage these effects in liquids such as SQ and HEX. The treatment is further complicated since estimates of δ_2 from IGC refer to concentrated solutions whereas more traditional methods are often applicable to dilute solutions and the relationship, if any, between these values is unclear.

Given the theoretical shortcomings of the solubility parameter concept, it is probably misguided to attempt an explanation of the results in strict thermodynamic terms. However, as previously mentioned, χ has been interpreted as an internal energy parameter and Pressure-Volume, or Equation of State, effects which are completely ignored in the solubility parameter treatment, will be more significant in small molecule systems where thermal expansion coefficients are generally greater. Hence χ for these systems may include contributions that are relatively insignificant in polymer studies. However, it is not clear why such a good linear relation is obtained for a variety of polymers. Further experimental work, especially on small molecule systems with more experimental results including more polar probes is necessary for a complete understanding of the effects involved in these systems.

Literature Cited

1. Orwoll, R.A. Rubber Chem. Technol. 1977, 50, 451.
2. Bonner, D.C. J. Macromol. Sci. Rev. Macromol. Chem. 1975, C13, 263.
3. Ashworth, A.J.; Price, G.J. Thermochim. Acta. 1984, 82, 161.
4. Lipson, J.E.G.; Guillet, J.E. In Developments in Polymer Characterization -3; Dawkins, J.V., Ed.; Applied Science Publishers: Barking, 1982.
5. Gray, D.G. Prog. Polym. Sci. 1977, 5, 1.
6. Smidsrod, O.; Guillet, J.E. Macromolecules 1969, 2, 272.
7. Ashworth, A.J.; Price, G.J. Macromolecules 1984, 17, 1090.
8. Newman, R.D.; Prausnitz, J.M. J. Phys. Chem. 1972, 76, 1492.
9. Deshpande, D.D.; Patterson, D.; Schreiber, H.P.; Su, C.S. Macromolecules 1974, 7, 630.
10. Olabisi, O.; Robeson, L.M.; Shaw, M.T. Polymer-Polymer Miscibility; Academic Press: London, 1979.
11. Su, C.S.; Patterson, D.D.; Schreiber, H.P. J. App. Polym. Sci. 1976, 20, 1025.
12. Prigogine, I. The Molecular Theory of Solutions; North Holland Publishing Co: Amsterdam, 1957.
13. Patterson, D. Macromolecules 1969, 2, 672.
14. Flory, P.J. Disc. Farad. Soc. 1970, 49, 7.
15. Sanchez, I; Lacombe, R. Macromolecules 1978, 11, 1145.
16. deGennes, P.G. Scaling Concepts in Polymer Physics; Cornell University Press: London, 1979.
17. Barton, A.F.M. Chem. Rev., 1975, 75, 731.
18. Harris, F.W.; Seymour, R.B. Eds. Structure-Solubility Relationships in Polymers; Academic Press: New York, 1977.

19. Hildebrand, J.H.; Scott, R.W.; Prausnitz, J.M. Regular and Related Solutions;, Van Nostrand: New York, 1972.
20. DiPaola-Baranyi, G; Guillet, J.E. Macromolecules 1978, 11, 228.
21. Bristow, G.M.; Watson, W.F. Trans. Farad. Soc. 1958, 54, 1731.
22. Braun, J.M.; Guillet, J.E. Macromolecules 1975, 8, 557.
23. DiPaola-Baranyi, G. Macromolecules 1982, 15, 622.
24. Merk, W.; Lichtenthaler, R.; Prausnitz, J.M. J.Phys.Chem. 1980, 84, 1694.
25. Fernandez-Berridi, M.J.; Guzman, G.M.; Iruin, J.J.; Elorza, J.M. Polymer 1983, 24, 417.
26. DiPaola-Baranyi, G.; Guillet, J.E.; Klein, J.; Jeberien, H.E. J. Chromatogr. 1978, 166, 349.
27. Ito, K.; Guillet, J.E. Macromolecules 1979, 12, 1163.
28. Lipson, J.E.G.; Guillet, J.E. in Solvent-Property Relations in Polymers; Seymour, R.B. and Stahl, G., Eds.; Pergamon: New York, 1982.
29. Braun, J.M.; Guillet, J.E. Adv. Polym. Sci., 1976, 21, 108.
30. Price, G.J.; Purnell, J.H.; Guillet, J.E. J. Chromatogr., 1986, 369, 273.
31. Brandrup, J.; Immergut, E.H., Eds.; Polymer Handbook; Wiley: New York, 1975.
32. Lipson, J.E.G.; Guillet, J.E. J. Polym. Sci. Phys. 1981, 19, 1199.
33. Conder, J.R.; Young, C.L. Physicochemical Measurement by GC; Wiley: Chichester, 1979.
34. Chien, C.F.; Laub, R.J.; Kopecni, M.M.; Smith, C.A. J. Phys. Chem. 1981, 85, 1864.
35. Harbison, M.; Laub, R.J.; Martire, D.E.; Purnell, J.H.; Williams, P.S. J. Phys. Chem., 1979, 83, 1262.
36. Ferreira, P.; Bastos, J.; Medina, A. J. Chem. Eng. Data 1987, 32, 25.
37. Perry, R.W.; Tiley, P.F. J. Chem. Soc. Farad. I., 1978, 74, 1655.
38. Hansen, C.M. J. Paint Technol. 1967, 39, 104.

RECEIVED September 29, 1988

Chapter 6

Gas and Vapor Adsorption on Polymeric Materials by Inverse Gas Chromatography

Bao Shiyao[1], S. Sourirajan[1], F. D. F. Talbot[1], and T. Matsuura[2]

[1]Department of Chemical Engineering, Industrial Membrane Research Institute, University of Ottawa, Ottawa, Ontario K1N 6N5, Canada
[2]Division of Chemistry, National Research Council of Canada, Ottawa, Ontario K1A 0R6, Canada

The adsorption data of different gases on various polymeric materials were obtained by using inverse gas chromatography (IGC) and compared with data available in the literature. An attempt was also made to obtain adsorption isotherms for a binary gas mixture. IGC offers a means to obtain gas adsorption data quickly. However, some improvements of the technique are necessary. Particularly, a high pressure IGC system must be developed to obtain adsorption data of gases under high pressures.
IGC was also applied to generate adsorption data for organic vapor on polymeric materials. The vapor-adsorption phase equilibrium for various binary mixtures of organic compounds was further calculated on the basis of adsorption data for individual vapors. These data are important in understanding vapor permeation through polymeric membranes, which occurs in the pervaporation process.

The adsorption of gas and vapor on polymeric materials is one of the factors governing gas and vapor permeation through polymeric membranes. For this reason, adsorption data have been determined for many polymeric films (1-6).

However, since conventional equilibrium absorption experiments can be time-consuming, an easier and simpler method is sought. This is particularly important when building a large data bank for the adsorption of gases and vapors on different polymeric materials. Such data may offer criteria for a preliminary screening of polymeric materials prior to the preparation of membranes for a given gas or vapor separation. Preferential adsorption of solute or solvent from the solution on the polymeric surface can be studied by using liquid chromatography (7). Furthermore, other aspects in the interfacial

0097–6156/89/0391–0059$06.00/0
Published 1989 American Chemical Society

properties of polymers at the polymer-solution interface have been
studied by liquid chromatography for a variety of polymers, and
results related to the permeation data of solution components in
reverse osmosis systems (8-14).

By analogy to the above technique, gas chromatography is
considered a useful tool to obtain data for gas and vapor adsorption
on polymeric surfaces. In contrast to liquid chromatography, the
general principle of the IGC technique is well established for the
characterization of polymeric materials; this technique called
inverse gas chromatography (IGC), enables the study of various polymeric properties, including interfacial properties (15-18).

The objective of this work was to demonstrate the feasibility of
IGC to generate data on the adsorption of gaseous and vaporous
adsorbates on the surface of polymeric materials. A precise measurement of the volume of dead space involved in the IGC system is
required to acquire adsorption data for weakly adsorbed gas
molecules. However, this problem is less serious for strongly
adsorbed vaporous adsorbates. Reflecting the difference in adsorption strength, Henry's law is generally applicable to gaseous
adsorbates, particularly for the low-pressure adsorption, while a
multi-layer adsorption isotherm is obtained for organic vapors. The
adsorption data obtained experimentally are further discussed in
relation to the gas and vapor permeation through polymeric membranes.
The drawbacks involved in this technique and the possibility for
improvement are also discussed.

Theory

The method of generating N_a (moles of adsorbed gas or vapor per unit
gram of polymer) versus p (partial pressure of gas or vapor) by IGC
is based on the method described by Mohlin and Gray (16). This
technique is the same as Elution by Characteristic Point (ECP) Method
described by Conder and Young (18).

According to the method N_a is given by

$$N_a = \frac{S_{locus} \cdot N}{m \cdot S_p} \tag{1}$$

where

$$S_{locus} = \int_0^h (d_a - d_u)dh \tag{2}$$

The area S_{locus} corresponds to the shadowed area shown in Figure 1.
Further, p is given by

$$p = \frac{NRT}{\gamma Sp} h \tag{3}$$

All symbols involved in Equations 1, 2 and 3 are defined in the
nomenclature.

The specific surface area of the polymer was determined using the BET approach. The BET equation may be written as

$$\frac{p/p_0}{N_a(1 - p/p_0)} = \frac{1}{N_{am}C} + \frac{C - 1}{N_{am}C}(p/p_0), \tag{4}$$

where p_0 is the saturation vapor pressure of the adsorbate vapor, and N_{am} is the amount adsorbed on the surface at the monolayer coverage. The constant C is related to the heat of adsorption. The quantities C and N_{am} can be determined from the slope and the intercept of the straight line obtained when $(p/p_0)/\{N_a (1 - p/p_0)\}$ is plotted against p/p_0 in the p/p_0 range 0.05 to 0.35 (16). To obtain the surface area of the polymer from the value of N_{am}, the area A_m that each adsorbate molecule covers must be known. This area can be calculated using the following equation (16),

$$A_m = 1.091 \times (\frac{M}{\rho \cdot N_0})^{2/3}, \tag{5}$$

where M and ρ are molecular weight and density of the adsorbate molecule, respectively, and N_0 is Avogadro's number, assuming that the molecular arrangement on the surface is the same as on a plane surface within the bulk of liquid. Then, the specific surface area of the polymer can be calculated as the product of N_{am} and A_m.

Materials and methods

In the IGC experiment, helium was used as a carrier gas. Pure gases of H_2, N_2, O_2, CO, CO_2, CH_4, C_3H_8, and C_2H_4; binary mixtures of CO_2/CH_4; and organic vapors of ethanol, 2-propanol, 1,4-dioxane, heptane, and octane were used as adsorbates. All gases were supplied by either Air Products, Inc. or by Matheson of Canada, Ltd. with purity of more than 99.9%. All organic compounds were of reagent grade. Cellulose acetate (CA-E398, supplied by Eastman Kodak Chemicals, Inc.), cellulose triacetate (CTA, supplied by Eastman Kodak Chemicals, Inc.), polyethersulfone (Victrex 200P, supplied by Imperial Chemical Industries), and cellulose (CE chromatography grade, supplied by Baker Chemical, Inc.) were the polymeric materials packed in the chromatography column. CA, CTA and CE were supplied in powder form by the manufacturers, while the Victrex was in pellet form. Pellets were crushed mechanically and sieved in the laboratory. Powders in the sieve range 38 to 53 μm were used as packing materials. Stainless steel chromatography columns with an inner diameter of 0.2295 cm were used. The column length and weight of the packing polymeric material are given in Table I for each polymer. Before adsorption experiments were started the column was flushed with a dry helium gas stream for approximately 12 hours to remove all traces of adsorbed gases from the previous experiment. Adsorption experiments were performed at a helium flow rate of 5 to 6 cm^3/min and an oven temperature of 35°C for gaseous adsorbates; a helium flow rate of 10.92 cm^3/min and an oven temperature of 24.3 to 27.3°C were used for organic vapors. The pressure drop through the column was kept below 2 kPa. The amount injected was 0.002 to 0.5 mL for gas samples, and 0.1 to 20 μL for organic liquid samples.

Table I. Characteristics of Chromatographic
Columns Used in This Study

Polymer	Column Length cm	Polymer Weight g	Polymer Volume cm^3	$[(V_R')]_u$ cm^3
CA-398	90.6	1.237	0.9515	3.5551
CTA	62.7	1.410	1.0905	2.2619
Victrex	60.8	1.330	0.9896	2.2842
Cellulose	69.8	0.772	–	–

Organic liquids were vaporized at 180°C and the vapor joined the
helium carrier gas stream immediately after injection. The
chromatography experiments were performed using either a
Spectraphysics SP 7100 Model or a Varian Aerograph Series 1400, both
equipped with a thermal conductivity detector.

Results and Discussion

Gas Adsorption. The precision in the value of d_u, the distance on
the chart corresponding to retention time (or volume) of the
unadsorbed gas, affects significantly the results obtained for S_{locus}
and consequently N_a, particularly when the adsorbate is in gaseous
form and only weakly adsorbed on the polymeric material. Therefore,
a careful measurement of d_u values was attempted in this work. The
retention volume of the unadsorbed gas, $[V_R']_u$, is related to d_u by

$$[V_R']_u = \frac{Q}{q} d_u \qquad (6)$$

which is equal to the sum of all dead spaces in the chromatographic
system. The dead space was obtained as the sum of the dead space I,
including those originating from the injector, the detector and the
connecting tube, and the dead space II, which is the space in an
chromatography column unoccupied by the polymer (Long, V.T.; Minhas,
B.S.; Matsuura, T.; Sourirajan, S. J. Colloid Interface Sci., in
press.). The dead space I was determined as (the retention volume
obtained when a chromatography column was replaced by an empty tube -
the volume inside the empty tube). The dead space II was determined
as (the volume inside an unpacked column - (polymer weight/polymer
density)).

Using d_u values obtained above, S_{locus} was determined by the
method described in the theoretical section and N_a calculated from
Equation 1. The value for p was calculated from Equation 3. The
results are shown in Figures 2, 3, and 4 for CA-398, CTA, and
Victrex, respectively, with regard to different gaseous adsorbates.
All the adsorption isotherms are almost linear with a slight
curvature in the range of the adsorbate gas pressure studied. The
only exception is hydrogen adsorption to CTA. The slope of the

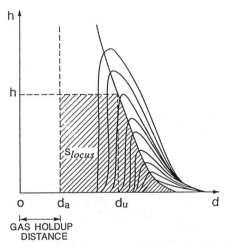

Figure 1. Superimposed chromatographic peaks for the calculation of area, S_{locus}. (Reproduced with permission from ref. 22. Copyright 1988 Academic Press.)

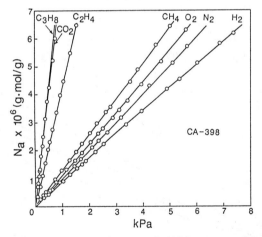

Figure 2. Adsorption isotherms of different gases on cellulose acetate 398 polymer at 35°C. (Reproduced with permission from ref. 22. Copyright 1988 Academic Press.)

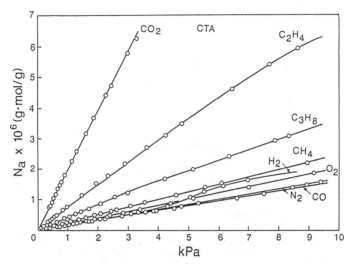

Figure 3. Adsorption isotherms of different gases on cellulose
triacetate polymer at 35°C. (Reproduced with permission from ref.
22. Copyright 1988 Academic Press.)

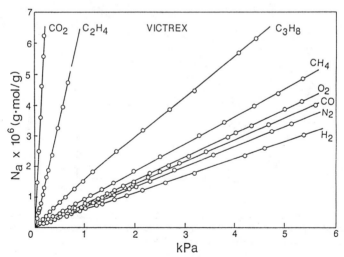

Figure 4. Adsorption isotherms of different gases on polyethersulfone Victrex polymer at 35°C. (Reproduced with permission from
ref. 22. Copyright 1988 Academic Press.)

Table II. Henry's Constants for Adsorption of Various
Pure Gases on CA-398, CTA, and Victrex Polymers at 35°C

| Polymer | Henry's Constant, H x 10^5 mol/g·kPa [a] | | | | | | | |
| | Adsorbate Gas | | | | | | | |
	CO_2	C_2H_4	C_3H_8	CH_4	O_2	N_2	CO	H_2
CA-398	0.752	0.416	0.871	0.109	0.109	0.0763	–	0.0866
CTA	0.188	0.0768	0.0404	0.0257	0.0197	0.0164	0.0157	0.0175
Victrex	2.855	0.657	0.157	0.0915	0.0742	0.0638	0.0706	0.0550

[a] Gas pressure below 100 kPa

adsorption isotherm is Henry's constant and designated here as H.
The numerical value for H was determined by applying linear
regression analysis to the data shown in Figures 2, 3, and 4. The
results of this analysis, as listed in Table II, show that Henry's
constant decreases in the following order:

$$CO_2 > C_2H_4 > C_3H_8 > CH_4 > O_2 > N_2, CO, H_2$$

for the polymers studied, except CA, where the order is

$$C_3H_8 > CO_2 > C_2H_4 > \ldots\ldots$$

The order in Henry's constant among different polymers, on the
other hand, depends on the adsorbate.

Gas Mixtures. Because of the importance of the separation of CO_2/CH_4
gas mixtures, it is interesting to compare the ratio H_{CO_2}/H_{CH_4}. The
ratios are 31.2, 5.9, and 7.31 for Victrex, CA, and CTA,
respectively, indicating the superiority of Victrex as membrane
material for CO_2/CH_4 gas separation from the perspective of gas
adsorption isotherms.
   Gas adsorption from the binary CO_2/CH_4 mixture to cellulose
acetate was studied at different compositions. For this study,
different volumes of gaseous mixtures of a given composition were
injected into the column. A chromatogram with two peaks, one for
CO_2 and the other for CH_4, was obtained for each injection and the
method illustrated in Figure 1 for determining S_{locus} and h was
applicable to both peaks. The latter S_{locus} and h values were used
to calculate N_a, p and Henry's constant $H=N_a/p$. As a result of this
calculation, H_{CO_2} and H_{CH_4} at different gas compositions were
obtained. In Figure 5, $X_{CO_2} H_{CO_2}$, $X_{CH_4} H_{CH_4}$, and $X_{CO_2} H_{CO_2} +$
$X_{CH_4} H_{CH_4}$ are illustrated as functions of X_{CO_2}, where X_i is the mole
fraction of component i. If the adsorption is completely ideal,

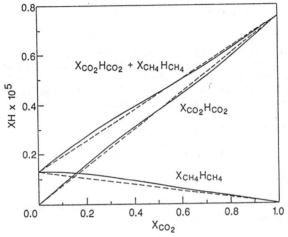

Figure 5. Adsorption of methane/CO$_2$ gas mixture on cellulose acetate 398 polymer at 35°C. (Reproduced with permission from ref. 22. Copyright 1988 Academic Press.)

these correlations should be linear, as illustrated in Figure 5 by the broken lines. The real adsorption curve is slightly nonlinear. Specifically, when the CO_2 mole fraction is less than 50% the amount of CO_2 adsorbed is less than the amount expected for ideal adsorption; however, it is greater than the amount expected for ideality when the mole fraction of CO_2 is more than 50%. On the other hand, methane adsorption is more than expected for ideality in the entire range of gas compositions of the binary mixture.

Vapor Adsorption. Ethanol and heptane vapor adsorption isotherms on cellulose are illustrated in Figure 6 for temperatures of 24.3°C and 26.5°C. The isotherms indicate multi-layer adsorption, which is typical for vapor adsorption on polymeric materials (16). Adsorption is sensitive to the temperature of both vapors and is decreased significantly by increasing the temperature from 24.3°C to 26.5°C. The saturation vapor pressures are also shown in the graph for both temperatures. The figure shows that the maximum vapor pressure on the adsorption isotherm is only 67% of the saturation vapor pressure for ethanol and 71% for heptane. In both adsorption isotherms, the amount of the adsorbed organic vapor increases steeply near the maximum vapor pressure included in the isotherm.

The isotherms for 1,4-dioxane, ethanol, heptane, and octane are shown in Figure 7. The temperatures at which the isotherm curves were obtained are 24.3°C for 1,4-dioxane, ethanol and heptane, and 25°C for octane. Though the temperature for octane is slightly higher than that for the other organic vapors, the difference in the temperature seems to be small enough to make the comparison of these adsorption isotherms meaningful. Furthermore, relative pressure (vapor pressure/saturation vapor pressure) was used for the pressure scale instead of the vapor pressure itself. The amount of adsorbed vapor decreases in the order 1,4-dioxane > ethanol > heptane > octane.

The order (1,4-dioxane > alcohol > hydrocarbon) is in agreement with the data reported by Mohlin and Gray (16). The solubility parameters (MPa) of cellulose and the organic vapors studied are:

Cellulose (49.3) > ethanol (26.0) > 1,4-dioxane (20.2) > heptane (15.1).

Therefore, the adsorption strength can be expected to be ethanol > 1,4-dioxane > heptane if the affinity between cellulose and organic vapor governs the adsorption strength. However, the experimental results show that 1,4-dioxane is more strongly adsorbed than ethanol. This is probably due to the higher boiling point of 1,4-dioxane (101.5°C) than that of ethanol (78°C) indicating that the condensibility of 1,4-dioxane vapor on the polymeric surface is higher than that of ethanol vapor.

Figure 8 shows the plot of $(p/p_0)/\{N_a (1 - p/p_0)\}$ versus p/p_0, the latter ratio being in the range proposed by Mohlin and Gray (16). The only exception is octane vapor for which the range of p/p_0 extends to 0.39. Excellent straight line relationships were obtained for all the organic vapors studied. Numerical values of A_m, N_{am}, and surface area are listed in Table III. All the specific surface area

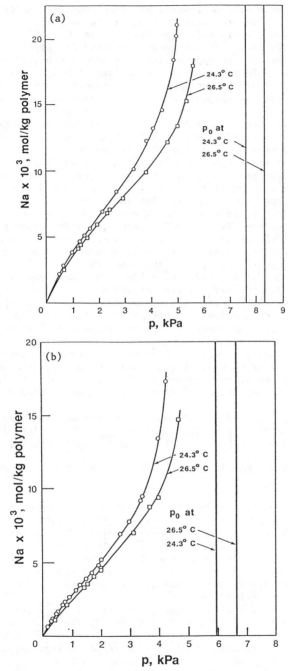

Figure 6. Adsorption isotherms of (a) ethanol and (b) heptane vapors at 24.3°C and 26.5°C on cellulose polymer.

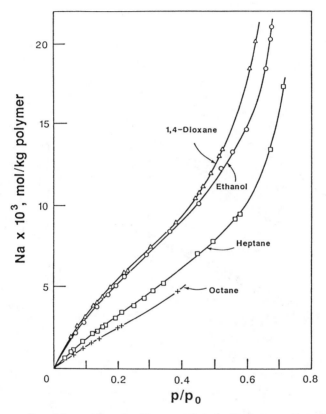

Figure 7.   Adsorption isotherms of 1,4-dioxane, ethanol, heptane, and octane on cellulose polymer.   Temperatures were 24.3°C, except for the octane isotherm, which was obtained at 25°C.

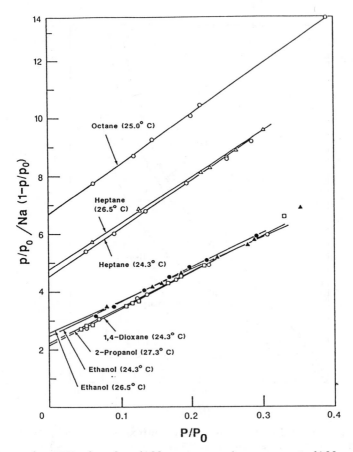

Figure 8. BET plot for different organic vapors at different
temperatures.

Table III. Specific Surface Area of Polymer
Measured for Different Organic Vapors [a]

Vapor	Temperature °C	A_m x 10^{20}, m^2	N_{am} x 10^{21}, $\dfrac{molecules}{kg \cdot polymer}$	Surface Area x 10^{-3}, m^2/kg
Ethanol	24.3	23.09	4.36	1.01
	26.5	23.13	4.48	1.04
2-Propanol	27.3	27.72	4.10	1.14
1,4-Dioxane	24.3	29.46	4.10	1.21
Heptane	24.3	42.7	2.85	1.21
	26.5	42.8	2.91	1.24
Octane	25.0	45.76	2.50	1.14

[a] polymer: cellulose

data are in the range 1.01 to 1.24 m^2/g, which are lower than the
values obtained by Mohlin and Gray (1.6 to 1.7 m^2/g). The specific
surface area data obtained above also indicate that cellulose is not
swollen significantly by these organic vapors. These surface areas
are particularly in contrast to 243 m^2/g, a value obtained for water
vapor in previous work (19). Such an enormous increase in surface
area is thought to be due to swelling of the cellulose in a water
vapor environment.

Vapor Mixtures. Phase equilibrium curves correlating the mole
fraction of component A in the adsorbed phase, $X_{A,ads}$, and that in
the gas phase, $X_{A,gas}$, were estimated for two binary mixtures of
liquid components A and B. The equilibrium phase diagram is shown in
Figure 9 for ethanol/heptane mixtures (9-a) and for 1,4-dioxane/
heptane mixtures (9-b) at 24.3°C. The method of generating these
equilibrium phase diagrams for the binary mixture of ethanol (A) and
heptane (B) is as follows. The mole fraction, $X_{A,ads}$, for a given
mole fraction $X_{A,gas}$ is calculated at a total vapor pressure p. The
partial vapor pressure of the component A in the vapor phase is
$pX_{A,gas}$, and that for the component B in the vapor phase is $pX_{B,gas}$ =
$p(1 - X_{A,gas})$. Then, using Figures 6a and 6b, the moles of adsorbed
vapor of component A per kilogram of polymer, N_{aA}, and that for the
component B, N_{aB}, can be obtained. Further, by using N_{aA} and N_{aB}
obtained above, $X_{A,ads}$ is calculated as $X_{A,ads} = N_{aA}/(N_{aA} + N_{aB})$.
Similarly, $X_{B,ads} = N_{aB}/(N_{aA} + N_{aB})$. This method is based on the
assumption that the adsorption of components A and B are completely
independent and additive, which may not always be true and therefore,
the phase equilibrium lines obtained above should be confirmed by
further experiments. However, the above method is effective in
determining the effect of the total vapor pressure on the phase
equilibrium diagram. Some interesting results have been obtained in
this respect.

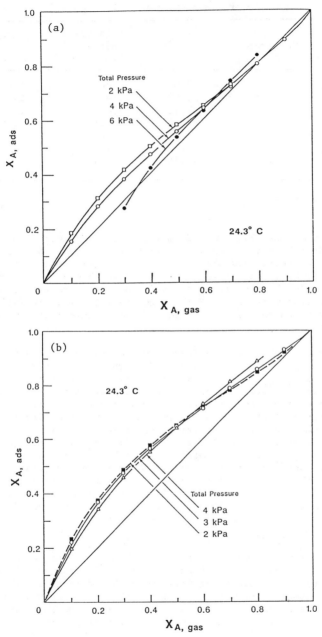

Figure 9. Mole fraction in the adsorbed phase, $X_{A, ads}$, versus
mole fraction in the vapor phase, $X_{A, gas}$, at 24.3°C.
   a) A - ethanol, B - heptane
   b) A - 1,4-dioxane, B - heptane

Figure 9-a shows that by increasing the total pressure, $X_{A,ads}$ (A is ethanol) in the high $X_{A,gas}$ range is increased, while the trend is reversed in the low $X_{A,gas}$ range. In fact, $X_{A,ads}$ may be lower than $X_{A,gas}$ in a low $X_{A,gas}$ range. The same trend is observed for the system 1,4-dioxane/heptane (Figure 9-b). The main feature of the adsorbed phase-vapor equilibrium depicted above seems reasonable, considering the adsorption isotherm curves for ethanol and heptane shown in Figure 6-a and 6-b.

For example, when the total pressure and $X_{A,gas}$ are high, the partial pressure of component A(ethanol) becomes high and, according to Figure 6-a, the amount of adsorbed vapor of the component A, N_{aA}, is large, since N_{aA} is found in the range where the curve is steeply increasing. However, the partial pressure of component B is low, since $X_{B,gas}$ is low and the amount of the adsorbed vapor of the component B is small, since N_{aB} is found in the range where N_{aB} increases with vapor pressure only at a modest rate. Therefore, at a high $X_{A,gas}$ value, a large $X_{A,ads}$ ($=N_{aA}/(N_{aA} + N_{aB})$) is expected at a high total pressure. For a small $X_{A,gas}$ value, the trend is reversed, resulting in a small amount of component A (ethanol) adsorption and a large amount of the component B (heptane) adsorption and consequently, a small $X_{A,ads}$ value. This explains why a high total pressure pushes up $X_{A,ads}$ values in the high $X_{A,gas}$ range and pushes down $X_{A,ads}$ values in the low $X_{A,gas}$ range.

A pervaporation system consists of equilibria at both sides of the membrane. One side of the membrane is in contact with the feed liquid mixture, while the other side is exposed to the permeate vapor at low pressure. It is considered that equilibria are established locally at both sides of the membrane. Adsorption equilibrium at the liquid-polymer interface must be established on the feed side, while an adsorption equilibrium at the vapor-polymer interface must be established on the permeate side. Further, both sides of the membrane are connected by liquid phase and gas phase diffusions of permeant molecules in the polymer. Therefore, adsorption equilibria at both liquid-polymer and vapor-polymer interfaces must be studied to fully discuss pervaporation phenomena. This aspect is neglected in many pervaporation papers. Although adsorption at the liquid-polymer interface can be studied by inverse phase liquid chromatography (20,21), this paper shows that adsorption at the vapor-polymer interface can be studied by IGC.

Conclusion

Inverse gas chromatography is a useful tool to study polymer-gas or polymer-vapor interfacial properties and, particularly, to generate adsorption isotherm data. There are several improvements required in this technique.
1. In this technique, it is intrinsically assumed that the presence of inert gas, such as helium, does not affect the adsorption of other gases and vapors. This assumption must be tested.
2. By the presently available technique, the upper limit of the gas pressure studied is only 100 kPa. In order to generate adsorption data at a higher gas pressure, a high pressure gas chromatographic technique must be applied.

3. When the vapor adsorption is strong, a large quantity of adsorbate liquid must be injected. The time required for injection and the subsequent vaporization causes an error in the retention time data. Low temperature operations are difficult for the same reason.
4. It is impossible to study the adsorption of the binary mixtures of organic vapors, since chromatograms of both components overlap. However, this problem may be solved by applying computer techniques for splitting the unresolved peaks.

Despite the limitations mentioned above, the chromatographic method is easy to handle and generates adsorption isotherm data quickly for a large number of combinations of polymers and gases or vapors. These data are important when membrane gas separation and pervaporation data are interpreted, and polymer materials are chosen for the preparation of membranes used for the above separation processes.

Acknowledgments

Thanks are due to Mr. V.T. Long and Dr. B.S. Minhas for the gas adsorption experiments. Issued as IMRI No. 3 and NRCC No. 28995.

Legend of Symbols

A_m = the area covered by each adsorbate molecule, m^2/molecule
C   = constant related to heat of adsorption
c   = molar concentration of adsorbate in the carrier gas stream, mol/m^3
d   = distance on the chromatography chart, m
H   = Henry's constant, mol/g polymer·kPa
h   = recorder response, m
M   = molecular weight, kg/k mol
m   = the amount of polymer packed in the column, kg
N   = the amount of adsorbate gas or vapor injected into the column, mol
N_a = the amount of adsorbed gas or vapor on unit weight of polymer, mol/kg
N_{am} = N corresponding to a monolayer coverage, mol/kg
N_o = Avogadro's number
p   = gas or vapor pressure, Pa
p_o = saturation vapor pressure, Pa
Q   = carrier gas flow rate, m^3/s
q   = recorder chart speed, m/s
R   = gas constant
S_p = the area of the adsorbate peak, m^2
S_{locus} = the area bounded by the common curve along the peak height maxima, the gas hold-up distance, and a given recorder response h, m^2
T   = temperature, K
$[V_R']$ = retention volume, m^3

Greek Letters

γ = ratio of the volume flow rate of the carrier gas to the recorder chart speed, m^2

ρ = density, kg/m^3

Literature Cited

1. Barrie, J.A. In Diffusion in Polymers; Crank, J.; Park, G.S., Eds.; Academic: New York, 1968; p 259.
2. Hauser, J.; Heintz, A.; Reinhardt, G.A.; Schmittecker, B.; Wesslein, M.; Lichtenthaler, R.N. Proceedings of Second International Conference on Pervaporation Processes in the Chemical Industry, Bakish, R., Ed.; Bakish Materials Corp.: NJ, 1987; p 15.
3. Neel, J.; Nguyen, Q.T.; Clement, R.; Francois, R. Proceedings of Second International Conference on Pervaporation Processes in the Chemical Industry, Bakish, R., Ed.; Bakish Materials Corp.: NJ, 1987; p 35.
4. Lloyd, D.R.; Meluch, T.B. In Materials Science of Synthetic Membranes; Lloyd, D.R., Ed.; ACS Symposium Series No. 269, American Chemical Society: Washington, DC, 1985; p 47.
5. Chuduk, N.A.; Eltekov, Yu.A.; Kiselev, A.V. J. Colloid Interface Sci. 1981, 84, 149.
6. Pusch, W.; Tanioka, A. Desalination 1983, 46, 425.
7. Matsuura, T.; Sourirajan, S. J. Colloid Interface Sci. 1978, 66, 589.
8. Matsuura, T.; Blais, P.; Sourirajan, S. J. Appl. Polym. Sci. 1976, 20, 1515.
9. Matsuura, T.; Sourirajan, S. Proceedings of the Sixth International Symposium on Fresh Water from the Sea, Delyannis, A. and E., Ed.: Athens, 1978; p 227.
10. Taketani, Y.; Matsuura, T.; Sourirajan, S. Proceedings of the Symposium on Ion Exchange: Transport and Interfacial Properties, 158th Electrochemical Society Meeting, Hollywood, Florida, Oct 5-10, 1980; p 88.
11. Matsuura, T.; Taketani, Y.; Sourirajan, S. Proceedings 2nd World Congress of Chemical Engineering, 1981, Volume IV; p 182.
12. Taketani, Y.; Matsuura, T.; Sourirajan, S. Sep. Sci. Technol. 1982, 17, 821.
13. Taketani, Y.; Matsuura, T.; Sourirajan, S. J. Electrochem. Sec. 1982, 129, 1485.
14. Taketani, Y.; Matsuura, T.; Sourirajan, S. Desalination 1983, 46, 455.
15. Huber, J.F.K.; Gerritse, R.G. J. Chromatgr. 1971, 58, 1937.
16. Mohlin, U.-B.; Gray, D.G. J. Colloid Interface Sci. 1974, 47, 747.
17. Braun, J.-M.; Guillet, J.E. In Advances in Polymer Science; Springer: Berlin, 1976; p 107.
18. Conder, J.R; Young, C.L. In Physicochemical Measurement by Gas Chromatography; John Wiley and Sons: New York, 1979; Chap. 9.

19. Matsuura, T.; Taketani, Y.; Sourirajan, S. Desalination 1983, 46, 455.
20. Farnand, B.; Talbot, F.D.F.; Matsuura, T.; Sourirajan, S. Fifth Canadian Bioenergy R & D Seminar, S. Hasnain, ed.; Elsevier: New york, 1984; p 195.
21. Matsuura, T.; Sourirajan, S.; Farnand, B.; Talbot, F.D.F. International Congress on Membranes and Membrane Processes, Tokyo, June 8-12: Tokyo, 1987; p 293.
22. Long, V. T.; Minhas, B. S.; Matsuura, T.; Sourirajan, S. J. Colloid Interface 1988.

RECEIVED November 2, 1988

Chapter 7

Thermodynamic Study of Water Sorption and Water Vapor Diffusion in Poly(vinylidene chloride) Copolymers

P. G. Demertzis and M. G. Kontominas

Department of Chemistry, University of Ioannina, GR 453 32, Ioannina, Greece

Inverse gas chromatography, IGC, has been used to study water sorption of two poly (vinylidene chloride-vinyl chloride) and poly (vinylidene chloride-acrylonitrile) copolymers, at temperatures between 20 and 50°C and low water uptakes. It was found that the specific retention volume of water increases with decreasing amount of water injected, increases dramatically with decreasing temperature and strongly depends on the type of copolymer. Thermodynamic parameters of sorption namely free energy, entropy, enthalpy of sorption and activity coefficient were calculated. Values were interpreted on the basis of an active site sorption model. Diffusion coefficients and activation energies values, calculated from the slopes of the Van Deemter curves are in general accordance with previously published values.

The sorption of water by synthetic and biological polymers is an important and extensive field (1). Polymers investigated include polystyrene (2), vinyl polymers (3, 4), cellulose derivatives (4, 5), collagen (6), starch (7), and various copolymers (8).

Sorption of water vapor by polymers is a diffusional process (9). The rate and extent of water sorption depends on the diffusion coefficient of water in the polymer, on the water/polymer interaction, and on the temperature.

Most of the available data on diffusion and diffusion coefficients of volatile liquids or gases in polymers have been obtained by static sorption experiments (10, 11), which are time consuming and require extensive data analysis. In recent years, inverse gas chromatography, IGC, was found to have wide utility in measuring sorption tendency and diffusion coefficients of gases and volatile liquids in molten polymers (12-17).

IGC enables rapid determination of thermodynamic parameters as well as

0097–6156/89/0391–0077$06.00/0
© 1989 American Chemical Society

diffusion coefficients. However, IGC is applicable only to solutes at infinite dilution.

In this paper, the water sorption of two commercially available vinylidene chloride copolymers is studied using IGC at low probe concentrations. The copolymers are a poly (vinylidene chloride-vinyl chloride) copolymer (Saran B) and a poly (vinylidene chloride-acrylonitrile) copolymer (Saran F). These copolymers are extensively used in the form of films, coatings, and film laminates in various industrial applications (for example, packaging of foods and pharmaceuticals) where their diffusion characteristics are of prime importance.

Calculation of Thermodynamic Parameters

In gas chromatography, the net retention volume, Vn, is given by

$$Vn = j \; \overset{*}{V} \; (tr - tf) \tag{1}$$

where j is the James and Martin compressibility factor, accounting for pressure drop along the chromatographic column; $\overset{*}{V}$ is the corrected carrier gas flow rate (mL/s); tr is the retention time in seconds of the water; and tf is the retention time of a non-interacting compound (air).

The specific retention volume, $V^o g$ defined as the net retention volume per unit weight of polymer, corrected to 273 K, is given by

$$V^o g = j . \overset{*}{V} \; (tr - tf) \frac{1}{Ws} \frac{273}{Tc} \tag{2}$$

where Ws is the polymer weight (g) and Tc is the column temperature (K).

The net retention volume can be expressed as

$$Vn = Kp.Vs \tag{3}$$

where Kp is the partition coefficient (defined as the ratio between the concentration of solute in the polymer and in the mobile phase, respectively, and Vs is the volume of the polymer.

Combining equations 1, 2, and 3 results in

$$Kp = \frac{V^o g . \rho s . Tc}{273} \tag{4}$$

where ρs is the polymer density.

The total partial molar enthalpy of sorption ($\Delta H_T^o = \Delta H_S^o$) is related to $V^o g$ through the Clausius-Clapeyron ([18]) equation:

$$\frac{d \; (\ln V^o g)}{dT} = \frac{\Delta H_S^o}{RT^2} \tag{5}$$

The total partial molar Gibb's free energy of sorption ($\Delta G^{O}_{T} = \Delta G^{O}_{s}$) is directly related to Kp by

$$\Delta G^{O}_{s} = -RT \ln Kp \tag{6}$$

The total partial molar entropy of sorption ($\Delta S^{O}_{T} = \Delta S^{O}_{s}$) is then calculated using

$$\Delta S^{O}_{s} = \frac{\Delta H^{O}_{s} - \Delta G^{O}_{s}}{T} \tag{7}$$

Finally, the activity coefficient, χ, of the polymer/water interaction can be calculated using

$$\Delta G^{O}_{s} = RT \ln P^{O} \tag{8}$$

where P^{O} is the saturation vapor pressure of pure water.

Calculation of Diffusion Coefficient

Van Deemter et al. (19) related peak broadening in a gas chromatographic column to column properties through equation 9:

$$H = A + \frac{B}{u} + Cu \tag{9}$$

where H is the theoretical plate height, u is the linear velocity of the carrier gas, and A, B, and C are constants independent of u.

Whereas A and B are related to instrument performance and gas phase spreading, C depends on a number of factors, including the diffusion coefficient of the probe molecule in the stationary phase. The constant C is given by

$$C = \left(\frac{8}{\pi^{2}}\right) \left(\frac{d}{Dp}\right) \frac{K}{(1+K)^{2}} \tag{10}$$

where d is the thickness of the stationary phase, Dp is the diffusion coefficient of the probe molecule, and K is the partition ratio given by

$$K = \frac{(tr - tf)}{tn} \tag{11}$$

In Equation 11, tr and tf are the retention times to peak maximum of the probe molecule and a non-interacting material used as a marker, respectively.

The determination of Dp involves the measurement of H at several relatively high flow rates, where the term B/u is negligible. The slope obtained in a plot of H versus u enables one to calculate Dp, since K is known in these experiments. The plate height, H, is determined from the eluted peaks displayed on a chart recorder by

$$H = \left(\frac{L}{5.54}\right) \left(\frac{W_{1/2}}{tr}\right)^{2} \tag{12}$$

where L is the column length, W1/2 the peak width at half the peak height, and tr is the retention time from injection to peak maximum.

Braun et al. (13) used these equations to calculate the diffusion coefficients of several materials in low density polyethylene. Millen and Hawkes (20), following Giddings (21), claimed that a value of 2/3 should be used as the constant in Equation 10 instead of $8/\pi^2$.

In accordance with the theory of activated diffusion, a plot of log diffusion coefficient (Dp) versus 1/T should produce a straight line of slope $-E_a/2.303R$

$$Dp = Do^{-E_a/RT} \tag{13}$$

where Do is a constant, E_a the activation energy for diffusion, and R is the gas constant.

Materials and Methods

The copolymers investigated in the present study were both supplied by Polysciences, Inc., U.S.A. The first was poly(vinylidene chloride-vinyl chloride, 80:20% w/w), having an average molecular weight ($\bar{M}_w$) of 9×10^4. The second was poly(vinylidene chloride-acrylonitrile, 80:20% w/w) of unknown molecular weight.

A Varian gas chromatograph Model 3700, equipped with a thermal conductivity detector, was used in all experiments.

The column parameters are described in Table I. The chromatographic parameters were as follows:

flow rate (mL/min)	:45
carrier gas	; nitrogen, high purity
temperature (^oC) : injection port	: 100
column	: 20 to 50
detector	: 220

The column packing material was prepared as follows (22). The polymers were dissolved in tetrahydrofuran and coated onto the inert support by slow evaporation of the solvent with gentle stirring and heating. After vacuum drying for approximately 48 h at 50^oC, the chromatographic supports were packed into the columns with the aid of a mechanical vibrator. The thickness of the stationary phase (d) was determined by

$$d = (1/3) \, W \, (\rho s/\rho p) \, \bar{r} \tag{14}$$

where W is the % loading of the polymer, ρs is the density of inert support, ρp is the density of polymer and r is the radius of inert support particles. The temperature range studied in the present work was above the T_g of both copolymers.

Distilled water was injected into the chromatographic column in volumes ranging from 1 to 20 μL.

Table I: Stationary Phase and Column Parameters

Polymer type	Inert support	Loading (%)	Polymer mass (g)	Column dimensions (length x O.D., cm)
Poly(vinylidene chloride- vinyl chloride, 80:20%)	Chromosorb WAW, DMCS, 60/80	5.55	0.544	200 x 0.63
Poly(vinylidene chrolide- acrylonitrile, 80:20%)	Chromosorb WAW, DMCS, 60/80	5.55	0.564	200 x 0.63

Results and Discussions

A sample chromatogram indicating sorption curve shape is given in Figure 1. It is clear that as amount of water injected increases the front profile of the peak becomes increasingly diffused while the rear profile remains almost vertical. This pattern is similar to that previously reported (24) and can be explained by some kinetic process such as partial penetration of the probe into the bulk of the polymer, a phenomenon which would cause peak broadening.

The effect of the amount water injected, mp on the specific retention volume for the two copolymers at various temperatures is shown in Table II. The above effect at T=20°C is characteristically shown in Figure 2.

Table II and Figure 2 show that V^0g increases with decreasing amount of water injected; increases dramatically with decreasing temperature; and depends strongly on type of copolymer.

The first two findings are indicative of an active site sorption process. Moreover, for a given T and mp, V^0g values for the poly(vinylidene-acrylonitrile) copolymer are significantly higher than those of the poly(vinylidene-/vinyl chloride) copolymer, indicating a stronger polymer/water interaction in the first system. This difference between the two copolymers can be explained in terms of the higher affinity of the nitrile group for water, as compared to the affinity of the chlorine groups.

Thermodynamic parameter values for both systems are given in Tables III and IV. It is evident that for both systems ΔG^0_s decreases with decreasing temperature, indicating a more favorable interaction between polymer and water at low temperatures.

It is also evident that at a given temperature, ΔG^0_s values for the poly(vinylidene-acrylonitrile) copolymer are somewhat lower than those of the poly(vinylidene-vinyl chloride) copolymer, indicating a relatively favorable polymer/water interaction in the first case. The explanation for this is analogous to that for V^0g.

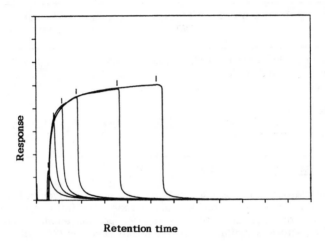

Retention time

Figure 1. Elution chromatogram of 1, 2, 4, 8, 12 and 20 µL water injected.

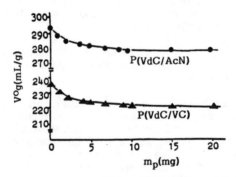

Figure 2. Specific retention volume as a function of the amount of water injected for the (▲) P(VdC/VC) and (●) P(VdC/AcN) copolymers at T=20°C.

Table II. Specific Retention Volume as a Function of the Amount of Water Injected for the Two P(VdC) Copolymers at Various Temperatures

a) P(VdC/VC)				b) P(VdC/AcN)			
	30°C	40°C	50°C		30°C	40°C	50°C
m_p (mg)	V^o_g (mL/g)	V^o_g (mL/g)	V^o_g (mL/g)	m_p (mg)	V^o_g (mL/g)	V^o_g (mL/g)	V^o_g (mL/g)
1	139.33	74.20	42.44	1	178.85	96.86	57.80
2	136.83	72.58	40.62	2	176.02	95.00	55.74
4	135.30	71.26	39.05	4	174.25	93.65	55.30
5	134.34	70.79	38.32	5	173.50	93.14	53.65
7	133.50	70.39	37.76	7	172.73	92.40	52.83
9	133.01	70.22	37.45	9	172.20	92.17	52.41
10	132.77	70.20	37.33	10	171.89	92.10	52.23
15	132.51	70.14	37.32	15	171.36	92.03	52.09
20	132.42	70.14	37.29	20	171.21	91.96	52.09

Table III. Thermodynamic Parameters of Interaction Between P(VdC/VC) and Water at Infinite Dilution (0.01 μg water)

T(K)	P^o(atm)	K_p	ΔG^o_s(Kcal/mole)	ΔS^o_s(cal/mole K)	$\chi \times 10^{-2}$
293	0.023	406.2	-3.50	-22.87	9.60
303	0.042	256.7	-3.34	-22.64	9.30
313	0.073	142.5	-3.08	-22.75	9.64
323	0.122	88.6	-2.88	-22.66	9.28

ΔH^o_s = -10.20 Kcal/mole

Table IV. Thermodynamic Parameters of Interaction Between P(VdC/AcN) and Water at Infinite Dilution (0.01 μg water)

T(K)	P^o(atm)	K_p	ΔG^o_s(Kcal/mole)	ΔS^o_s (cal/mole K)	$\chi \times 10^{-2}$
293	0.023	510.4	-3.63	-22.76	7.49
303	0.042	329.7	-3.49	-22.48	7.24
313	0.073	170.9	-3.20	-22.68	7.50
323	0.122	113.6	-3.04	-22.48	7.23

ΔH^o_s = -10.30 Kcal/mole

ΔS^o_s values for both systems are almost constant, indicating that there is probably no change in the degree of randomness of the system within the temperature range studied.

The enthalpy of sorption, a measure of the energy of attachment of sorbed molecules to the polymer, increases with increasing amount of water injected (Figure 3).

Values of ΔH^o_s close to -10.0 Kcal/mol (Figure 3) correspond to a strong polymer/water interaction and support the active site hypothesis. According to this hypothesis the water initially introduced is tightly bound onto certain irregularities in the polymer structure (active sites) ($\Delta H \simeq -10.2$ Kcal/mol). Once these active sites are covered, additional water molecules are held more loosely by the polymer ($\Delta H \simeq -9.5$ Kcal/mol). This value is close to the heat of condensation of water (-10.5 Kcal/mol) and one can expect sorption on sites as well as condensation to take place at higher water uptakes.

Both ΔH^c_s and ΔS^c_s values are in agreement with literature values of -7.8 Kcal/mole and -20.4 cal/mole K, respectively (8), and -8.5 Kcal/mole (23) for a similar vinyl chloride–vinyl acetate copolymer, P(VC/VAc).

Finally, χ values (mole fraction) given in Tables III and IV are indicative of strong attractive forces between polymer and water. Although the activity coefficients do not change significantly with temperature, their trend is to increase with increasing temperature, pointing toward a decrease in degree of interaction.

Diffusion coefficients and activation energies for the two copolymers calculated from the slopes of Figures 4 and 5 are given in Table V.

Table V. Diffusion Coefficients and Activation Energies for Water in the Two P(VdC) Copolymers between 20°C and 50°C

Polymer	Temperature (°C)	Dp ($cm^2 . s^{-1} \times 10^{-8}$)	Ea (Kcal/mole)
P(VdC/VC)	20	1.63	
	30	2.00	2.75
	40	2.19	
	50	2.55	
P(VdC/AcN)	20	0.75	
	30	1.20	5.19
	40	1.50	
	50	1.75	

Diffusion coefficients are of the same order of magnitude for both copolymers. An average value of 2.0×10^{-8} cm^2/s corresponds well to a value of 2.38×10^{-8} cm^2/s for a PVC homopolymer (3); however, it differs significantly from an average value of 17×10^{-6} cm^2/s reported for a P(VC/VAc) copolymer (8) at similar temperatures. Differences are probably due to the presence of plasticizer, (25% dioctyl phthalate) which causes inter-segmental bond weaking and therefore, results in diffusion coefficient increase. Apparently, diffusion coefficients increase with temperature increases. Activation energy values are in general agreement with those of similar copolymers like P(VC/VAc) for which reported values were between 2.2 and 7.7 Kcal/mole (8) at same temperatures.

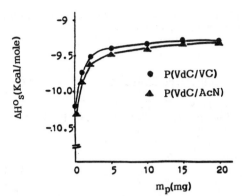

Figure 3. Heat of sorption of both P(VdC/VC) -water (●) and P(VdC/AcN) - water (▲) interaction at 20 to 50°C as a function of the amount of water sorbed.

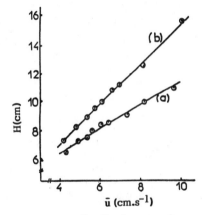

Figure 4. Van Deemter curves for the two copolymers: (a) P(VdC/VC) and (b) P(VdC/AcN).

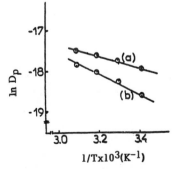

Figure 5. Arrhenius plots of diffusion coefficients for water in the two copolymers (a) P(VdC/VC) and (b) P(VdC/AcN).

In conclusion, IGC has shown to be a powerful tool for the evaluation of the thermodynamics of water sorption by polymers at low water uptake levels, as well as for the determination of diffusion coefficients.

Literature Cited

1. Franks, F. Water: A Comprehensive Treatise; Plenum Press: New York 1975; Chpts. 5-7.
2. Garcia-Fierro, J.L. Eur. Polym. J. 1985, 21(8) 753.
3. Tikhomirov, B.P.; Hopfenberg, H.B.; Stannett, V. and Williams J.L. Die Makromol.Chem. 1968, 118; 177.
4. Long, F.A.; Thompson, L.J. J. Polym. Sci. 1955, 15, 413.
5. Aspler, J.S.; Gray, D.G. J. Polym. Sci. 1983, 21, 1675.
6. Coelho, U.; Miltz, J.; Gilbert, S.G. Macromolecules 1979, 12 284.
7. Carrillo, P. Ph.D. Thesis, Rutgers University, New Brunswick, NJ 1988.
8. Kumins, C.A.; Rolla, C.J.; Rotoman, J. J. Phys. Chem 1957, 61, 1290.
9. Crank, J. The Mathematics of Diffusion; Clarendon Press; Oxford 1975.
10. Duda, J.L.; Ni, Y.C.; Vrentas, J.S. J. Appl. Polym Sci. 1979, 23, 947.
11. Ju, S.T.; Duda, J.L.; Vrentas, J.S. Ind. Eng. Chem. Prod Res. Dev. 1981, 20, 330.
12. Gray, D.G.; Guillet, J.E.; Macromolecules 1973 6, 223.
13. Braun, J.M.; Poos, S.; Guillet, J.E. J. Polym. Sci. Polym Lett. Ed 1970, 14, 257.
14. Kong, J.M.; Hawkes, S.J. J. Chromatogr. Sci. 1978, 14, 279.
15. Edwards, T.J.; Newman, J. Macromolecules 1977, 10, 609.
16. Tait, P.J.T.; Abyshihada, A.M. J. Chromatogr. Sci. 1979, 17, 219.
17. Hu, D.S.; Han, C.D.; Stiel, L.J. J. Appl. Polym. Sci. 1987, 33, 551.
18. Kiselev, A.V.; Yashin, Y.J. Gas Adsorption Chromatography; Plenum Press: New York 1969.
19. Van Deemter, J.J.; Zuiderweg, F.J.; Klikenberg, A. Chem. Eng. Sci. 1966, 5, 271.
20. Millen, W; Hawkes, S.J.; Polym. Lett. 1977, 15, 463.
21. Giddings, J.C. Anal. Chem 1963, 135, 439.
22. DiPaola-Baranyl, G.; Guillet, J.E. Macromolecules 1978, 11(1), 228.
23. Doty P. J. Chem. Phys. 1946, 14, 244.
24. Gray, D.G.; Guillet, J.E. Macromolecules 1972, 5(3), 316.

RECEIVED September 29, 1988

Chapter 8

Solute Diffusion in Polymers by Capillary Column Inverse Gas Chromatography

Dominique Arnould and Robert L. Laurence

Department of Chemical Engineering, University of Massachusetts, Amherst, MA 01003

Inverse Gas Chromatography (IGC) has been developed for accurate measurement of diffusion coefficients in polymer-solvent systems at conditions approaching infinite dilution of the volatile component. Recently, the technique has been extended to coated capillary columns. In this paper, the applicability of capillary columns to IGC measurements is reviewed, and results of a study on the effect of penetrant size and configuration on diffusion in polymethyl methacrylate are presented. Measurements of diffusion coefficients have been made over a range of temperatures. The effect of solvent size on the activation energy, E_D, is examined in the limit of zero mass fraction of solvent. The available diffusion data do not allow discrimination between the conflicting theories describing the variation of E_D with solvent size: the ceiling value hypothesis and the hypothesis based on free-volume theory. The extent of segmental motion for large and sufficiently flexible solvents and its effect on diffusion has also been investigated. Finally, the measurement of methanol diffusion coefficients near the glass transition temperature are discussed.

The mass transfer of low molecular weight molecules in concentrated polymers plays an important role in many polymer processing steps. During the formation of the polymer, the rate of polymerization is sometimes influenced by the diffusion of low molecular weight species. After the reaction step, a devolatilization process is usually used to remove volatile residuals from the polymer via solvent diffusion. Many other processes, such as the distribution of additives in polymer, involve the diffusion of low molecular weight molecules in concentrated polymer solutions.

Diffusivity data are available only for a limited number of polymer-solvent systems. This paper describes research that has led to the development of the use of capillary column inverse gas chromatography (IGC) for the measurement of diffusion coefficients of solute molecules in polymers at infinite dilution. The work has resulted in a precise, rapid technique for the diffusion measurements that circumvents the many problems attendant to classical sorption methods and packed column IGC methods. Initial results of the program appeared in two recent publications ([1],[2]). Some of the material introduced in those papers is discussed here to present background for

0097–6156/89/0391–0087$06.00/0
© 1989 American Chemical Society

the discussion, but new data is also presented on polymethyl methacrylate that allows careful examination of the merits of a free-volume theory for diffusion.

Theory of Diffusion
The theories that describe diffusion in concentrated polymer solutions are approximate in nature. Among them, only one seems sufficiently developed to offer a good description of mass transfer in polymer-solvent systems : the *free-volume theory* of diffusion. Though it affords good correlative success, it needs further testing.

A brief review is presented of the theories describing transport processes in binary solutions of an amorphous, uncross-linked polymer and low molecular weight solutes. At present, there exists no theory capable of describing diffusion in polymer-solute systems over the entire concentration range. No general theory has been formulated to describe diffusional transport under conditions where viscoelastic effects are important. However, methods have been developed to anticipate conditions under which anomalous effects can be expected (3-9). This brief review is limited to the theories applicable for concentrated polymer solutions under conditions where the classical diffusion theory holds.

At present, all these theories are approximate, since all attempts to derive them using molecular mechanics have been largely unsuccessful, because there is a large number of degrees of freedom in describing concentrated polymer solutions. Among these approximate theories, such as those developed by Barrer (10), DiBenedetto (11), and van Krevelen (12), the free-volume theory of diffusion is the only theory sufficiently developed to describe transport processes in concentrated polymer solutions.

Cohen-Turnbull Free-Volume Theory. The original free-volume work was developed to describe transport in liquids (13-15). Bueche (16) considered volume fluctuations to analyze polymer segmental mobility. Cohen and Turnbull (17,18) gave the free-volume theory its first theoretical basis by developing an expression relating the self-diffusion coefficient to the free volume for a liquid of hard spheres. In their free-volume model for molecular transport in dense fluids, the voids that allow the molecular transport are represented by a random distribution of the free volume in the material. Naghizadeh (19) proposed a modified version of the Cohen-Turnbull theory by considering a redistribution energy for the voids. Macedo and Litovitz (20) broadened the Cohen-Turnbull theory by taking into account attractive forces in addition to the repulsive forces, as well as the availability of free space. Chung (21) placed the Macedo-Litovitz analysis on a solid theoretical basis by deriving it using statistical-mechanical arguments. Turnbull and Cohen (22) improved their free-volume model by taking into account the variable magnitude of the diffusive displacement.

Application to Polymer-Solvent Systems. Fujita (23) was the first to use the free-volume theory of transport to derive a free-volume theory for self-diffusion in polymer-solvent systems. Berry and Fox (24) showed that, for the temperature intervals usually considered (smaller than 200^0C), the theories that consider a redistribution energy for the voids gives results similar to those of the theories that assume a zero energy of redistribution for the free volume available for molecular transport. Vrentas and Duda (5,6) re-examined the free-volume theory of diffusion in polymer-solvent systems and proposed a more general version of the theory presented by Fujita. They concluded that the further restrictions needed for the theory of Fujita are responsible for the failures of the Fujita theory in describing the temperature and concentration dependence of diffusion coefficients for binary mixtures of small molecules and amorphous polymers. Paul (25) used the Cohen-Turnbull theory to develop a model for predicting solvent self-diffusion coefficients in polymer-solvent solutions. A major advantage of the Paul model is that it contains only three parameters. In contrast to the Vrentas-Duda model, no diffusion data are needed for the evaluation of the parameters. Paul's theory is expected to give good predictions, however, only for polymer concentrations less than 0.9. Vrentas, Duda, and Ling (9,26) compared their free-volume theory for self-diffusion with Paul's, both conceptually and experimentally, and concluded that the Vrentas-Duda version gives better agreement with existing data.

There are some important considerations in the Duda-Vrentas theory that bear some examination; for example, the effect of the solvent size on diffusional behavior, and the behavior of the diffusion process near the glass transition.

Effect of the Solvent Size. The effect of solvent size and geometry is reflected in the apparent activation energy for diffusion, E_D, for diffusion in the limit of zero solvent concentration defined by

$$E_D = R T^2 \left[\frac{\partial \ln (D_1)}{\partial T} \right] \qquad (1)$$

Here, D_1 is the self-diffusion coefficient of the solvent. In the limit of zero solvent concentration, it is equal to the mutual-diffusion coefficient D.

The variation of E_D with the solvent size has been described by two conflicting theories :

i) Kokes (27) and Meares (28) proposed that the activation energy approaches a ceiling value as the molecular size of the solvent increases. For solvents larger than a polymeric jumping unit, the activation energy takes the ceiling value. The movement of such solvents are controlled by the motion of polymer molecules.

ii) The Vrentas-Duda free-volume theory predicts that the activation energy increases indefinitely as the size of the solvent jumping unit increases. However, these authors have indicated that the solvent jumping unit size does not necessarily increase when the size of the entire molecule increases, since sufficiently large and flexible solvent molecules exhibit segmental motion. In that case, the effective size of the solvent molecule is smaller than its total size. Vrentas and Duda (8, 29) examined this problem using published data for a variety of solvents diffusing in polystyrene and concluded in favor of their interpretation of the effect of the solvent size on the activation energy. However, though the work of Vrentas and Duda has examined a broad size range for penetrants, more data are needed to establish conclusively the influence of the solvent molecular size and geometry on its diffusional behavior.

Glass transition. Vrentas and Duda (7) have shown that the free-volume theory predicts a step change in the activation energy at the glass transition temperature. This step change is a function of the solvent size, the glass transition temperature, the free-volume characteristics of the polymer, and the change in the thermal expansion coefficient at the transition. At present, the paucity of data precludes the testing of the theory. A key problem with measuring diffusion data below the glass transition temperature is to control the amount of free volume that is frozen in the polymer. This free volume is a function of the mechanical and thermal history of the polymer as it passes through the glass transition. Hence, the thermal history of the polymer must be controlled to allow comparison of diffusivity data of different solvents.

Diffusion Measurements.

Conventionally, diffusion coefficients for solutes in molten polymers are determined by gravimetric sorption/desorption experiments. A sample of known weight and shape is placed on a sensitive balance and exposed to a constant concentration of solute. From the weight gain of the sample versus time, a diffusion coefficient is calculated. Crank and Park (30) and Crank (31) reviewed the method in detail. Disadvantages of the technique include the relatively long time required for a single measurement, the difficulty of maintaining a constant, simple geometry for the molten sample, and the need to use solute concentrations large enough to produce measurable weight changes. These constraints limit the precision of measurements at conditions approaching infinite dilution of the solute and for systems where the diffusion coefficient is strongly concentration dependent.

Inverse Gas Chromatography. A technique that promises to circumvent many of the problems attendant to gravimetric sorption experiments is Inverse Gas Chromatography (IGC). Until recently, all reported applications of IGC to the measurement of diffusion coefficients have used packed chromatographic columns in which the stationary phase is supported on a granular substrate. Equations similar to those developed by van Deemter et al. (32) are used to calculate the stationary phase diffusion coefficient from the spreading of the elution profile. The equation developed by van Deemter is commonly written as

$$H = A + B/V + CV \qquad (2)$$

where H is the height equivalent to a theoretical plate (HETP), and V is the mean velocity of the carrier gas. The constants A, B, and C represent the contributions of axial dispersion, gas phase molecular diffusion, and stationary phase mass transfer resistances toward broadening of the peak. The equation is only valid for describing the elution of symmetric peaks, which requires that mass transfer resistances be small, but not negligible. From plate theory, it can be shown that for a column producing Gaussian-shaped peaks, the HETP is related to the peak width, or variance, by the following (32):

$$H = L \{\sigma_t^2/t_r^2\} \quad = L/t_r^2 \{W_{1/2}/2.335\}^2 \qquad (3)$$

where L is the column length, σ_t^2 is the variance of the peak, t_r is the retention time of the peak, and $W_{1/2}$ is the width of the peak at half-height.

For the case in which all mass transfer resistance is due to diffusion in the stationary phase and the stationary phase is uniformly distributed on the surface of a uniform spherical packing, the constant C is related to the solute diffusivity (32) by

$$C = (8/\pi^2)(T^2/D_p)(K/\varepsilon)[1 + (K/\varepsilon)]^{-2} \qquad (4)$$

where D_p is the diffusion coefficient in the stationary phase, τ is the film thickness, K is the partition coefficient, and ε is the ratio of the stationary phase volume to the gas phase volume.

These results may be used to determine diffusivity from experimental data as follows: Solute elution curves are obtained for a range of flow rates. From measurements of peak width, a plot of H versus V is prepared. At sufficiently high flow rates, the second term on the right side of Equation 2 becomes negligible, and the plot is linear. From the measured slope, D_p is calculated using Equation 4. It is presumed in the analysis that diffusion in the stationary phase is Fickian and that the diffusion coefficient is concentration independent.

This technique is especially suited to the study of solutes with low diffusivities. When the solute diffusivity is small, diffusion within the stationary phase is the dominant process determining the shape of the elution profile, so that the contribution of other processes are neglected more readily.

Clearly, further improvements in the reliability and accuracy of the IGC method depend on the development of more suitable columns to support the stationary phase. Several authors have speculated that the use of capillary columns, or open tube columns would eliminate some of the concerns cited above, and would be advantageous for IGC applications (33-35). The principal attraction of a capillary column is the possibility of achieving more uniform dispersal of the polymeric phase. Ideally, the polymer would cover the wall as a uniform annular film. Such a geometrical configuration would simplify modelling of the transport processes within the column, and improve the inherent reliability and accuracy of IGC measurements.

Following the ideas developed by Guillet and his co-workers, a method using Capillary Column Inverse Gas Chromatography (CCIGC) was developed (1,2) to measure diffusion coefficients in polymer-solvent systems at conditions approaching infinite dilution of the volatile component.

The polymer is deposited as a uniform annular coating in a glass capillary column. A solute is injected into an inert carrier gas that flows through the column. The elution curve of the sample is then used with a model to determine the solute activity and diffusivity in the stationary phase. A detailed description of the equipment and the experimental procedure is given by Pawlisch (36). It is of value to present the model used to describe the process. The description provided by Pawlisch (36) , given below, indicates how the model was developed.

Model . The model used was developed by Macris . The assumptions made are the following:

(1) the column is a straight cylindrical tube;
(2) the system is isothermal;
(3) the carrier gas is treated as an incompressible fluid;
(4) the carrier flow is steady laminar flow with a parabolic velocity profile;
(5) the polymer stationary phase is homogeneous;
(6) the polymer film is constant in thickness;
(7) the polymer film thickness is significantly less than the radius of the column;
(8) the axial diffusion in the stationary phase is negligible;
(9) the carrier gas is insoluble in the polymer;
(10) the absorption isotherm is linear;
(11) no surface adsorption occurs at the polymer-gas interface or the polymer-column interface;
(12) no chemical reaction occurs between the sample gas and the polymer;
(13) diffusion coefficients are concentration independent over the range of interest; and
(14) the inlet concentration profile is modelled as an impulse function.

A modified version of this model was developed for a nonuniform polymer film (2). With these assumptions, the continuity equations for the gas and polymer phase may be written as

$$\frac{\partial c}{\partial t} + 2V[1 - (r/R)^2]\frac{\partial c}{\partial z} = D_g\left[\frac{1}{r}\frac{\partial}{\partial r}\left(\frac{r\partial c}{\partial r}\right) + \frac{\partial^2 c}{\partial z^2}\right] \tag{5}$$

and

$$\frac{\partial c'}{\partial t} = D_p\left[\frac{1}{r\partial r}\left(\frac{r\partial c'}{\partial r}\right)\right] \tag{6}$$

where c and c' are the gas phase and stationary phase solute concentrations, D_g and D_p are the gas phase and stationary phase diffusion coefficients for the solute, z and r are the axial and radial coordinates, and V is the mean velocity of the carrier gas. Appropriate initial and boundary conditions for the problem are:

$$c(r,z,t) = c'(r,z,t) = 0 \qquad \text{at } t = 0, z > 0; \tag{7}$$
$$c(r,z,t) = \delta(t)c_0 \qquad \text{at } z = 0; \tag{8}$$
$$c(r,z,t) = c'(r,z,t)/K \qquad \text{at } r = R; \tag{9}$$
$$D_g\,(\partial c/\partial r) = D_p\,(\partial c'/\partial r) \qquad \text{at } r = R; \tag{10}$$
$$\partial c/\partial r = 0 \qquad \text{at } r = 0; \tag{11}$$
$$\partial c'/\partial r = 0 \qquad \text{at } r = R + \tau; \tag{12}$$

where δ (t) is the Dirac delta function, c_0 is the strength of the inlet impulse, K is the partition coefficient, R is the radius of the gas-polymer interface, and τ is the thickness of the polymer film.

The problem stated above is sufficiently complex that a closed-form analytical solution in the time domain has not been found. For most purposes, the details of the radial distribution of solute are unimportant, and a description of the longitudinal dispersion of solute in terms of a local mean concentration (that is, radially averaged) will suffice. The most mathematically convenient mean concentration is an area-averaged concentration, defined as

$$c = \left\{\int_0^R rc\,dr\right\} \Big/ \left\{\int_0^R r\,dr\right\} \tag{13}$$

Application of this definition to Equations 5, 6, and 7, making use of the boundary conditions given by Equations 8 to 10, yields the following:

$$\frac{\partial c}{\partial t} + \frac{4V}{R^2}\int_0^R (1 - (r/R)^2)\frac{\partial c}{\partial z}\,rdr - D_g\frac{\partial^2 c}{\partial z^2} = \frac{2\,D_P}{R}\frac{\partial c'}{\partial r}\bigg|_{r=R} \tag{14}$$

$$c = c' = 0 \qquad \text{at } t = 0 \quad ; \quad c = \delta(t)\,c_0, \text{ at } z = 0 \tag{15}$$

The equations derived from radial averaging still contain the local concentration as a variable. To proceed further, approximations are developed to relate the local concentration to the area-averaged concentration. The approach used in earlier models

(39,40) was to define a new variable, Δc, which describes the deviation of the local concentration from the mean concentration,

$$c(r,z,t) = \mathbf{c}(z,t) + \Delta c(r,z,t) \tag{16}$$

When the chromatographic peak is dispersed, the radial variation in the gas phase concentration is expected to be small, so that $\Delta c \ll \mathbf{c}$. This approximation is used with Equations 14 to 16 to obtain an approximate solution in terms of $\mathbf{c}$ and its derivatives. That result may then be used to eliminate the local concentration from Equations 8 and 14.

A variety of models are generated by using different assumptions to obtain an approximate solution for $\Delta c(r,z,t)$. A plug-flow model follows from the simplest approximation for $\Delta c(r,z,t)$, namely that $\Delta c(r,z,t) = 0$. This is equivalent to stating that the radial gas phase concentration gradients are sufficiently small that $c(r,z,t) = \mathbf{c}(z,t)$. Substitution of this approximation into Equation 13, and evaluation of the integral yields a plug-flow model for the gas phase:

$$\frac{\partial \mathbf{c}}{\partial t} + V\frac{\partial \mathbf{c}}{\partial z} - D_g \frac{\partial^2 \mathbf{c}}{\partial z^2} = \frac{2D_P}{R} \left.\frac{\partial c'}{\partial r}\right|_{r=R} \tag{17}$$

The interfacial equilibrium boundary condition becomes

$$\mathbf{c}(z,t) = c'(r,z,t)/K \qquad \text{at } r = R \tag{18}$$

The assumption of plug flow is valid as long as radial transport processes occur more rapidly than those processes that create radial concentration variations. A more detailed discussion of the implications of this assumption has been given by Edwards and Newman (38).

The problem is now made dimensionless by introducing the following variables:

$$y = (\,\mathbf{c}L/c_0V) \quad ; \quad x = (z/L) \quad ; \quad \eta = (r-R)/\tau \tag{19}$$
$$q = c'L/c_0KV \quad ; \quad \theta = Vt/L$$

where L is the length of the column. The transport equations describing the elution process then can be expressed in dimensionless form as:

$$\frac{\partial y}{\partial \theta} + \frac{\partial y}{\partial x} = \frac{\gamma \partial^2 y}{\partial x^2} + \frac{2}{\alpha\beta^2} \left.\frac{\partial q(0)}{\partial \eta}\right|_{\eta=0} \tag{20}$$

$$\frac{\partial q}{\partial \theta} = \frac{1}{\beta^2} \frac{\partial^2 q}{\partial \eta^2} \tag{21}$$

where $\alpha = R/K\tau$; $\gamma = D_g/VL$; $\beta^2 = \tau^2 V/D_pL$. (Equation 21 was simplified, recognizing $\tau \ll R$.) The initial and boundary conditions that remain, written in dimensionless form, are

$$y = q = 0 \text{ at } \theta = 0 \;;\; y = \delta(\theta) \quad \text{at } x = 0 \tag{22}$$

$$y = q \qquad \text{at } \eta = 0 \; ; \; \partial q / \partial \eta = 0 \; \text{at } \eta = 1 \tag{23}$$

This pair of coupled linear equations may be solved using Laplace transforms. Solution of Equations 20-23 yields

$$Y(s,x) = \exp(1/2\gamma) \, \exp [-(1/2\gamma)(1 + 4\gamma\Psi(s))^{1/2})x] \tag{24}$$

where $\Psi(s) = s + (2s^{1/2}/\alpha\beta) \tanh(\beta s^{1/2})$.
At the exit of the column, where $x = 1$, the solution can be written as

$$Y(s,1) = \exp(1/2\gamma) \, \exp [-(1/2\gamma)(1 + 4\gamma\Psi(s))^{1/2}] \tag{25}$$

While relatively benign in appearance, this transform is difficult to invert analytically. The inversion scheme of Kubin (39) and Kucera (40), which uses a Hermite polynomial series expansion, is too cumbersome to be of any practical use. The coefficients are difficult to obtain algebraically and the series converges slowly.

Parameter Estimation. The model for the chromatographic experiment presented above describes a pulse response experiment: the elution curve is the response of the system (that is, the chromatogram) to an input disturbance, while the Laplace transform given by Equation 25 is the transfer function for the system. Methods for obtaining transfer function parameters from system response experiments are well developed and generally fall into one of four categories: time domain fitting, method of moments, Laplace domain fittting, and Fourier domain fitting (41,42). A discussion of the merits of each method for IGC applications is presented by Pawlisch (36). Each technique affords a simpler representation of this model, which can be used to affect parameter estimation.

In determining the partition coefficient and the solute diffusion coefficient in the stationary polymer phase, either moment analysis or Fourier domain fitting was used. The two techniques are described below.

Moment analysis. A technique for obtaining analytical information on the elution curve described by Equation 25 is to make use of the moment generating property of Laplace transforms (41). It is readily shown that the various moments of the real-time concentration profile are related to the transform solution by the following:

$$\mu_k = (-1)^k (L/V)^k \lim_{s \to 0} \frac{d^k Y(s)}{ds^k} \tag{26}$$

where

$$\mu_k = \int_0^\infty t^k c(t) \, dt \Big/ \int_0^\infty c(t) \, dt \tag{27}$$

The normalized moments are used to calculate central moments, which are frequently more meaningful in characterizing a distribution:

$$\mu_k^* = \int_0^\infty (t - \mu_1)^k \, c(t) \, dt \Big/ \int_0^\infty c(t) \, dt \tag{28}$$

The following equations were derived from Equation 30 for the first and second central normalized moments (35):

$$\mu_1 = \left(1 + \frac{2}{\alpha} \right) t_c \tag{29}$$

$$\mu_2^* = \left[\frac{4}{3} \frac{\beta^2}{\alpha} + 2\gamma \left(1 + \frac{2}{\alpha} \right)^2 \right] t_c^2 \tag{30}$$

with

$$\alpha = \frac{R}{\tau K} \quad ; \quad \beta^2 = \frac{\tau^2 V}{D_p L} \tag{31}$$

$$\gamma = \frac{D_g}{V L} \quad ; \quad t_c = \frac{L}{V} \tag{32}$$

The dimensionless mean retention time, μ_1/t_c, is independent of the carrier gas velocity and is only a function of the thermodynamic properties of the polymer-solute system. The dimensionless variance, μ_2^*/t_c^2, is a function of the thermodynamic and transport properties of the system. The first term of Equation 30 represents the contribution of the slow stationary phase diffusion to peak dispersion. The second term represents the contribution of axial molecular diffusion in the gas phase. At high carrier gas velocities, the dimensionless second moment is a linear function of velocity with the slope inversely proportional to the diffusion coefficient.

Fourier domain fitting. The Fourier transform of the experimental elution curve is calculated. The parameters α and β are then determined using a fitting procedure in the Fourier domain that is equivalent to a least-squares criterion in the time domain. With Fourier domain estimation, model parameters are chosen to minimize the difference between the Fourier transforms of experimental and theoretical elution curves. The Fourier transform of a bounded, time varying response curve, f(t), is defined as

$$G(\omega) = \int_{-\infty}^{\infty} f(t) \, e^{-i\omega t} dt \tag{33}$$

If f(t) = 0 for t < 0, then substitution of s = iω into the above equation yields the definition of the Laplace transform. Thus, the Fourier transform is obtained from the Laplace transform solution (Equation 25) by the substitution s = iω. The Fourier transform of an experimental elution curve is calculated at discrete values of ω by numerical integration of Equation 33.

The best criterion for minimizing the difference between the theoretical and experimental transform is not immediately obvious. In the time domain, a least-squares criterion is usually preferred (22); that is, parameters are selected that minimize the least-squares objective function

$$I = \int_{-\infty}^{\infty} \{f_t(t) - f_e(t)\}^2 \, dt \tag{34}$$

where $f_e(t)$ is the experimental response curve, and $f_t(t)$ is the theoretical response curve.

From Parseval's theorem, a least-squares criterion in the time domain is equivalent to the least-squares criterion in the Fourier domain (23):

$$I = \frac{1}{2\pi} \int_{-\infty}^{\infty} \{R_e(\omega) - R_t(\omega)\}^2 = \{I_e(\omega) - I_t(\omega)\}^2 \, d\omega \tag{35}$$

where $R(\omega)$ denotes the real part of the Fourier transform, $I(\omega)$ denotes the imaginary part of the Fourier transform, and the subscript indicates an experimental or theoretical transform. Minimization of this function with respect to the unknown model parameters results in the best least-squares approximation of the experimental elution curve in the time domain. The details of the procedure were presented by Pawlisch (36).

Validity of the Method. The feasibility of the method was demonstrated (36) by measuring diffusion coefficients and thermodynamic partition coefficients for benzene, toluene, and ethylbenzene in polystyrene between 110°C and 140°C. The measured values of the activity coefficient and diffusion coefficient were in agreement with data collected using other experimental techniques. More recently, Pawlisch, Bric, and Laurence (2) further demonstrated the utility of Fourier fitting in the analysis of elution data.

Experimental Program

The current interest is the examination of the consequences of free-volume theory on the effect of the solvent size on diffusional behavior, and the behavior of the diffusion process near the glass transition. Clearly, these two problems are interrelated. The experimental data needed to investigate both are accurate diffusivity-temperature data for a series of solvents that covers a wide range of molecular sizes. The series of solvents used should include solvents of large molecular size, incapable of segmental motion. Some recent work is reported here using polymethyl methacrylate, an amorphous polymer that can be studied over a wide temperature range.

The IGC technique is ideally suited for this study. Its two main advantages are its speed and precision of measurement. Once a column is prepared, a change of solvent and/or temperature is effected rapidly. The precision of the technique is associated with the fact that measurements are conducted at conditions approaching infinite dilution. Hence, the structure of the polymeric glass is not affected by the solvent. This is particularly important for measurements at temperatures close to the glass transition temperature.

Materials. A commercial grade polymethyl methacrylate (PMMA, PRD-41) was obtained from the Rohm and Haas Company. The weight-average molecular weight of the polymer and polydispersity, as determined by gel permeation chromatography, were 200,000 and 2.35, respectively. The solvents used (methanol, acetone, methyl acetate, ethyl acetate, propyl acetate, benzene, toluene, and ethylbenzene) were spectroscopic grade products obtained from Aldrich Chemical Company.

Column Preparation. PMMA column was prepared using the procedure described by Pawlisch (36). The film thickness was inferred from the values of diffusivities measured with a second PMMA column whose thickness was measured using a destructive characterization technique. The column used in this study had an axial length of 17.78 m, an inner radius of 367 μm, and a film thickness of 5.3 μm. This column was used to measure diffusivities for the PMMA/solvent systems in a temperature interval from 70°C to 160°C. For each temperature, measurements were taken at three different carrier gas flow rates. Replicate measurements were made at each flow rate, and replicate elution curves were obtained at some of the conditions to evaluate reproducibility.

Procedure. The apparatus and general procedures of the capillary IGC experiment are described elsewhere (1,2,36). Each measurement was conducted at three different carrier gas flow rates (between 2 to 20 cm/s). For each experiment, an estimate of α and β was obtained using moment analysis and used as an initial guess for the Fourier domain fitting. The values of β^2 at the three different carrier gas flow rates were plotted versus $1/t_c$. Using equation (31), τ^2/D was estimated from the slope of β^2 versus $1/t_c$, using a linear least-squares.

Data Acquisition. The original Apple II computer used by Pawlisch (36) was replaced by a Macsym 120 (Analog Devices). The microcomputer was used to record, store, and display the detector output signal. The primary improvement was derived from the larger memory in the computer (15,000 points in contrast to 1024 in the old system). A sampling frequency as rapid as 20 s^{-1} can be attained in contrast to 5 s^{-1} with the Apple II system. This has increased the accuracy and the precision of the data analysis. Firstly, more data points facilitate the determination of the baseline and the peak. Secondly, the fitting procedure could be improved as a result of a greater density of experimental points in the steep region of the elution curve. The data acquisition procedure is discussed in detail by Pawlisch (36). The code was rewritten for the Macsym 120 software. A graphics subroutine was added to plot the experimental elution curve and the theoretical curve obtained by the Fourier domain fitting.

Data Analysis. After each experiment, the data files were processed in two steps. The first step consisted of correcting the raw data for baseline offset and calculating the lower order moments of the elution curve. This was achieved with an interactive graphics routine that allows estimation of the baseline and the range of the elution peak. The second step was determining the Fourier domain estimation of the parameters (2). The value of α and β determined by the moment analysis were used as initial estimates.

Results and Discussion

Overview. The capillary column IGC technique was used to determine the partition coefficients and diffusion coefficients of a number of solvents (methanol, acetone, methyl acetate, ethyl acetate, propyl acetate, benzene, toluene, and ethylbenzene) in poly(methyl methacrylate). Measurements below the glass transition temperature were obtained for the PMMA/methanol system.

Diffusion Above the Glass Transition.   Diffusivity data are presented in Table I.   The logarithm of the retention volume, V_g^0, for methanol is presented in Figure 1 as a function of the reciprocal temperature. The retention volume is related to the partition coefficient by the following equation:

$$V_g^0 = \frac{273.2 \text{ K}}{\rho_p \text{ T}} \tag{36}$$

where ρ_p is the density of the polymer.

Table I. Summary of the Diffusion Data

Solvent	Temperature	D
	(K)	$(m^2/s) * 10^{12}$
Methanol	433.2	12.9
	423.2	7.33
	412.9	4.21
	403.7	2.54
	393.6	1.40
	383.3	0.743
	373.6	0.441
	363.3	0.256
	353.0	0.118
	343.7	0.072
Acetone	443.5	2.86
	433.3	1.50
	423.2	0.688
	413.0	0.249
	404.0	0.116
Methyl Acetate	443.0	2.92
	432.8	1.55
	423.7	0.720
	413.5	0.291
Ethyl Acetate	443.1	1.53
	432.7	0.790
	423.6	0.337
Propyl Acetate	443.0	0.899
	432.7	0.395
	423.5	0.167
Benzene	443.0	0.818
	432.9	0.345
	423.9	0.152
Toluene	443.2	0.664
	433.0	0.260
	423.8	0.107
Ethyl Benzene	443.3	0.442
	432.8	0.164
	423.7	0.070

As noted above, the mean retention volumes determined from the mean elution times measured at three different flow rates and were invariant with flow rate. Linearity in the logarithm of V_g-versus-$1/T$ plot is present above and below the glass transition temperature with a change of slope at the glass transition temperature. Using this plot, the glass transition temperature appears to be near 111°C. Measured by

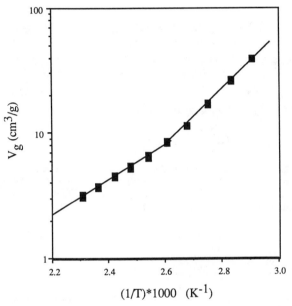

Figure 1 .   Retention volume versus temperature for the PMMA/methanol system.

differential scanning calorimetry, the glass transition temperature was 115 to 120°C. A glass transition temperature of 115°C is used in the following analysis.

The diffusivity data were analyzed using the Vrentas-Duda version of the free-volume theory. The basic equation describing the solvent and temperature dependence of the diffusion coefficient in the limit of zero mass fraction above the glass transition temperature is given by the expression

$$D_1 = D_{01} \exp\left(-\frac{E^*}{RT}\right) \exp\left(-\frac{\gamma \hat{V}_2^* \xi / K_{12}}{K_{22}+T-T_{g2}}\right) \tag{37}$$

where D_{01} is a pre-exponential factor; E^* is the critical energy per mole needed to overcome attractive forces, R is the universal gas constant; T is the absolute temperature; T_{g2} is the glass transition temperature of the polymer; ξ is the ratio of the critical molar volume of a solvent jumping unit to the critical molar volume of the jumping unit of the polymer, and K_{22} and $\gamma \hat{V}_2^*/K_{12}$ are related to the WLF constants of the polymer, C_1^g and C_2^g, by the following expression

$$\gamma \hat{V}_2^*/K_{12} = 2.303 \, C_1^g \, C_2^g \quad ; \quad K_{22} = C_2^g \tag{38}$$

The assumptions and restrictions of the free-volume theory, as well as the significance of its parameters, are discussed in detail by Vrentas and Duda (5,6). For temperatures close to the glass transition temperature, the diffusion process is free-volume dominated and the energy term can be absorbed in the pre-exponential term. Equation 36 becomes

$$D_1 = D_{01} \exp\left(-\frac{\gamma \hat{V}_2^* \xi / K_{12}}{K_{22} + T - T_{g2}}\right) \tag{39}$$

The apparent activation energy for diffusion in the limit of zero solvent concentration is expressed as

$$E_D = R \, T^2 \left[\frac{\partial \ln (D_1)}{\partial T}\right] = \frac{RT^2 (\gamma \hat{V}_2^* \xi / K_{12})}{(K_{22}+T-T_{g2})^2} \tag{40}$$

In Figures 2 to 4 are presented plots of the logarithm of D versus $(1/[K_{22}+T-T_{g2}])$, using $K_{22}-T_{g2} = -308°K$. Although the temperature range was limited, it can be concluded that the free-volume theory satisfactorily describes the temperature dependence of the diffusivity data except for the PMMA/methanol system. For this system, an Arrhenius plot is shown in Figure 5. The activation energy for diffusion seems to be independent of the temperature, although a slight curvature can be observed. For each PMMA/solvent system, the quantities D_{01} and $(\gamma \hat{V}_2^* \xi)/K_{12}$ determined using a least-squares regression, are presented in Table II. For the PMMA/methanol system, only the first three points above T_{g2} were used.

Following Vrentas and Duda (43) for solvent molecule moving as a single unit, the critical amount of hole-free volume per mole necessary for a solvent molecule

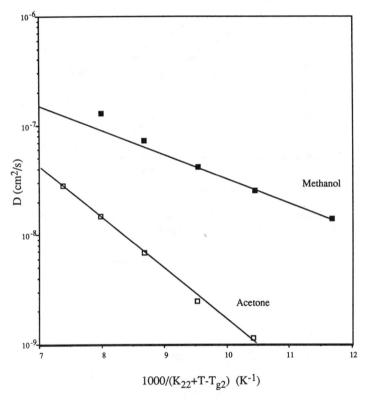

Figure 2.   Free volume correlation for diffusion as a function of temperature for methanol and acetone.

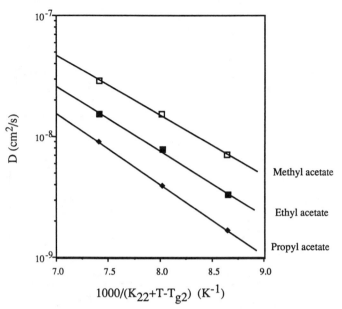

Figure 3 .   Free volume correlation for diffusion as a function of temperature for methyl acetate, ethyl acetate, and propyl acetate in PMMA.

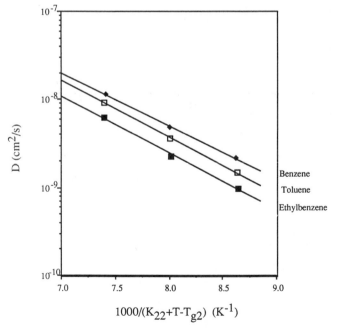

Figure 4 .   Free volume correlation for diffusion as a function of temperature for benzene, toluene, and ethyl benzene in PMMA.

Table II. Size Effects on Diffusion in Poly(Methyl Methacrylate)

Solvent	Molar Volume at 0°K (cm³/mole)	$\gamma V*_2\xi/K_{12}$ (K)	D_{01} (m²/s)	ξ
Methanol	30.8	510	0.0537	0.19
Acetone	58	1075	0.782	0.39
Methyl Acetate	63.3	1122	1.20	0.41
Ethyl Acetate	77.8	1215	1.28	0.45
Propyl Acetate	92.3	1346	1.93	0.49
Benzene	70.4	1377	2.18	0.50
Toluene	84.4	1473	3.51	0.54
Ethyl Benzene	98.5	1473	2.30	0.54

to jump can be taken as equal to the occupied volume of the liquid, defined as the molar volume of the liquid solvent at 0°K, $\tilde{V}_1^0(0)$, the following equation can be written :

$$\frac{\gamma \hat{\tilde{V}}_2^* \xi}{K_{12}} = \frac{\gamma \hat{\tilde{V}}_2^*}{K_{12}} \left(\frac{\tilde{V}_1^0(0)}{\tilde{V}_2^*} \right) \tag{41}$$

The parameters γ, and K_{12} are independent of the solvent. Thus, for a molecule moving as a single unit, $(\gamma \hat{\tilde{V}}_2^* \xi)/K_{12}$ should be a linear function of the molar volume of the liquid solvent at 0°K, $\tilde{V}_1^0(0)$. An average value of the solvent molar volume at 0°K was calculated using the methods of Sudgen and Biltz (Table II). A graph of $(\gamma \hat{\tilde{V}}_2^* \xi)/K_{12}$ versus the solvent molar volume at 0°K is shown in Figure 6. For solvents expected to move as a single unit (methanol, acetone, methyl acetate, benzene, toluene), the linear relationship between $(\gamma \hat{\tilde{V}}_2^* \xi)/K_{12}$ and $\tilde{V}_1^0(0)$, is reasonably represented by the experimental data. The straight line on figure 6 was obtained by using a least-squares method constrained to pass through the origin for the methanol, acetone, methyl acetate, benzene, and toluene data (solid squares on figure 6). The data presented in Figure 6 suggest that molecules of ethyl acetate, propyl acetate, and ethyl benzene move segmentally in PMMA. These data, obtained with diffusivity-temperature measurements in a single PMMA sample, reinforce the conclusions drawn by Vrentas and Duda (44) for Poly(methyl acrylate) (PMA), and allow the conclusion to be in favor of the free-volume interpretation of the effect of solvent size on the activation energy.

Diffusion Below the Glass Transition. The free-volume theory can also be used to analyze the influence of the glass transition temperature on the diffusivity of the PMMA/methanol system. The use of the free-volume theory both above and below the glass transition is discussed by Vrentas and Duda (7). According to these authors, below the glass transition temperature, Equation 39 becomes:

$$D_1 = D_{01} \exp\left(- \frac{\gamma \hat{\tilde{V}}_2^* \xi / \lambda K_{12}}{K_{22}/\lambda + T - T_{g2}} \right) \tag{42}$$

The parameter λ represents the character of the change, which can be attributed to the glass transition. For $\lambda = 1$, the equilibrium liquid, assumed above T_{g2}, is also

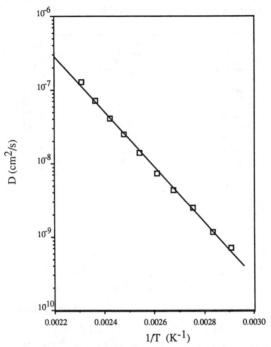

Figure 5 .   Arrhenius representation of the dependence with temperature of
the diffusion coefficients for the PMMA/methanol system.

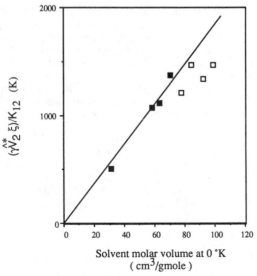

Figure 6 .   Representation of the dependence of the free-volume parameters with
molar volume of the solvent in PMMA.

realized below T_{g2}. At the other extreme, if $\lambda = 0$, the specific hole-free volume at any temperature is equal to the specific hole-free volume at T_{g2}. A rigorous definition of λ in terms of the free-volume parameters is given in the original paper (7).

A plot of the logarithm of $D(T)/D(T_{g2})$ versus $1/T$ for the PMMA/methanol system is shown in Figure 7. The data indicate that, for this system, the λ is slightly greater than 0.5, Vrentas and Duda predicted that an upper bound for λ for PMMA was 0.41. Consequently, the experimental value obtained for λ seems too large. However, it is important to note that figure 7 was constructed using $(\gamma\hat{V}_{2}^{*}\xi)/K_{12} = 510°K$, and that the value of this parameter has a great impact on the estimate of the value of λ.

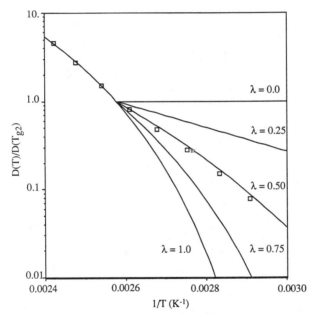

Figure 7 .   Diffusion of methanol in PMMA through the glass transition.

Acknowledgments

The support of the National Science Foundation, The Rohm and Haas Company, Mobil Chemical Company, and CUMIRP at the University of Massachusetts is gratefully acknowledged.

Literature Cited

1. Pawlisch, C. A.; Macris A.; Laurence, R. L. Macromolecules, 1987, 20, 1564.
2. Pawlisch, C. A.; Bric, J. A.; Laurence, R. L., Macromolecules, 1988, 21, 1685.
3. Alfrey, G.; Chem. Eng. News, 1965, 43, (41), 64.
4. Hopfenberg, H. B.; Frisch, H.L. J. Polym. Sci. 1969, B7, 405.

5. Vrentas, J. S.; Duda, J. L. J. Polym. Sci., Polymer Phys. Ed. 1977, 15, 403.
6. Vrentas, J. S.; Duda, J. L. J. Polym. Sci., Polymer Phys. Ed. 1977, 15, 417.
7. Vrentas, J. S.; Duda, J.L. J. Appl. Polymer Sci. 1978, 22, 2325.
8. Vrentas, J. S.; Duda, J. L.; Liu, H. T. J. Appl. Polymer Sci., 1980, 25, 1793.
9. Vrentas, J. S.; Duda, J. L.; Ling, H. C., J. Polym. Sci., Phys. Ed. 1985, 23, 275.
10. Barrer, R. M. J. Phys. Chem. 1957, 61, 178.
11. DiBenedetto, A. T. J. Polymer Sci. 1963, A1, 3477.
12. Van Krevelen, D. W. Properties of Polymers, Elsevier: Amsterdam, (1972).
13. Doolittle, A. K. J. Appl. Phys. 1951, 22, 1471.
14. Doolittle, A. K. J. Appl. Phys. 1952, 23, 236.
15. Doolittle, A. K.; Doolittle, D. B. J. Appl. Phys. 1957, 28, 901.
16. Bueche, F. J. Chem. Phys. 1953, 21, 1850.
17. Cohen, M. H.; Turnbull, D. J. Chem. Phys. 1959, 31, 1164.
18. Turnbull, D.; Cohen, M. H. J. Chem. Phys. 1961, 34, 120.
19. Naghizadeh, J. J. Appl. Phys. 1964, 35, 227.
20. Macedo, P.B.; Litovitz, T. A. J. Chem. Phys. 1965, 42, 245.
21. Chung, H. S. J. Chem. Phys. 1966, 44, 1362.
22. Turnbull, D.; Cohen, M. H. J. Chem. Phys. 1970, 52, 3038.
23. Fujita, H. Fortschr. Hochpolym-Forsch, 1961, 3, 1.
24. Berry, G.C.; Fox, T. G. Adv. Polymer Sci. 1968, 5, 261.
25. Paul, D. W. J. Polym. Sci. 1983, 21, 425.
26. Vrentas, J. S.; Duda, J. L.; Ling, H. C.; Hou, A. C. J. Polym. Sci.,Phys. Ed. 1985, 23, 289.
27. Kokes, R. J.; Long, F. A. J. Am. Chem. Soc. 1953, 75, 6142.
28. Meares, P. J. Polymer Sci. 1958, 27, 391.
29. Vrentas, J. S.; Duda, J. L. J. Appl. Polymer Sci., 1986, 31, 739
30. Crank, J., Park, G. S. Diffusion in Polymers Academic Press: New York, 1968.
31. Crank, J. The Mathematics of Diffusion Clarendon Press: Oxford, 1975; Chapter 10.
32. van Deemter, J. J.; Zuiderweg, F. J.; Klinkenberg, A. Chem. Eng. Sci. 1956, 5, 271.
33. Conder, J. R.; Young, C. L. Physicochemical Applications of Gas Chromatography John Wiley and Sons: New York ,1979.
34. Lichtenthaler, R. N.; Liu, D. D.; Prausnitz, J. M. Macromolecules 1974, 7, 565.
35. Macris, A., M. S. Thesis, University of Massachusetts, Amherst, 1979.
36. Pawlisch, C. A., Ph. D. Dissertation, University of Massachusetts, Amherst, 1985.
37. Jennings, W. Gas Chromatography with Glass Capillary Columns, 2nd ed., Academic Press: New York, 1980.
38. Edwards, T. J.; Newman, J. Macromolecules 1977, 10, 609.
39. Kubin, M., Coll. Czech. Chem. Commun., 1965, 30, 1104 .
40. Kucera, E. J. Chromatog. 1965, 19, 237.
41. Douglas, J. M. Process Dynamics and Control. Volume 1: Analysis of Dynamic Systems, Prentice-Hall Inc.: Englewood Cliffs, New Jersey , 1972.
42. Ramachandran, P. A; Smith, J. M. Ind. Eng. Chem. Fund. 1978, 17, 148 .
43. Vrentas, J. S.; Duda, J. L. J. Applied Polymer Sci. 1977, 21, 1715.
44. Vrentas, J. S.; Duda, J. L. J. Polymer Sci., Pol. Phys. ed. 1979, 17, 1085 .

RECEIVED September 1, 1988

POLYMER BLEND CHARACTERIZATION

Chapter 9

Thermodynamics of Polymer Blends by Inverse Gas Chromatography

G. DiPaola-Baranyi

Xerox Research Centre of Canada, 2660 Speakman Drive, Mississauga, Ontario L5K 2L1, Canada

IGC was used to determine the thermodynamic miscibility behavior of several polymer blends: polystyrene-poly(n-butyl methacrylate), poly(vinylidene fluoride)-poly(methyl methacrylate), and polystyrene-poly(2,6-dimethyl-1,4-phenylene oxide) blends. Specific retention volumes were measured for a variety of probes in pure and mixed stationary phases of the molten polymers, and Flory-Huggins interaction parameters were calculated. A generally consistent and realistic measure of the polymer-polymer interaction can be obtained with this technique.

The concept of blending two or more polymers to obtain new polymer systems is attracting widespread interest and commercial utilization. Blending provides a simpler and more economical alternative for obtaining polymeric systems with desired properties, as compared to the synthesis of new homopolymers. This growing demand for polymer blends has generated a need for a better understanding of the thermodynamics of miscibility and phase separation in polymer systems. This in turn has generated tremendous interest in techniques that can be used to characterize the thermodynamics of polymer-polymer systems.

The usefulness of inverse gas chromatography for determining polymer-small molecule interactions is well established ([1,2]). This method provides a fast and convenient way of obtaining thermodynamic data for concentrated polymer systems. However, this technique can also be used to measure polymer-polymer interaction parameters via a ternary solution approach ([3]). Measurements of specific retention volumes of two binary (volatile probe-polymer) and one ternary (volatile probe-polymer blend) system are sufficient to calculate χ_{23}', the Flory-Huggins interaction parameter, which is a measure of the thermodynamic

0097–6156/89/0391–0108$06.00/0

miscibility of two polymers. IGC has been used to study a
variety of blends. Some of these include polystyrene-
poly(dimethyl siloxane) ($\underline{4}$), polystyrene-poly(vinyl methyl ether)
($\underline{5},\underline{6}$), poly(methyl acrylate)-poly(epichlorohydrin) ($\underline{7}$),
poly(vinylidene fluoride)- poly(ethyl acrylate) ($\underline{8}$),
poly(ε-caprolactone)-poly(vinyl chloride) ($\underline{9},\underline{10}$), and
poly(dimethyl siloxane)- polycarbonate ($\underline{11}$).
    This paper reviews the application of IGC in determining
interaction parameters for three polymer blend systems:
polystyrene-poly(n-butyl methacrylate) (PS-PnBMA), polystyrene-
poly(2,6-dimethyl-1,4-phenylene oxide) (PS-PPO), and
poly(vinylidene fluoride)-poly(methyl methacrylate) (PVF_2-PMMA)
($\underline{12}$-$\underline{14}$). In each case, a generally consistent and realistic
measure of the polymer-polymer interaction is obtained.

Materials and Methods

Materials. All solutes were chromatographic quality or reagent
grade and were used without further purification. The
polystyrene samples (PS: M_W = 110,000, M_W/M_n <1.06; PS_L: M_n =
1709, M_W/M_n <1.06) were obtained from Polysciences and Pressure
Chemical Co., respectively. Poly(n-butyl methacrylate) (PnBMA:
M_W = 320,000, M_n = 73,500) was obtained from Aldrich. The
poly(2,6-dimethyl-1,4-phenylene oxide) sample was obtained from
General Electric Co. (M_W = 69,000; M_W/M_n = 2.1). Poly(vinylidene
fluoride), Kynar 881, was supplied by Pennwalt Corp., and
poly(methyl methacrylate), Acrylite H-12, was supplied by the
American Cyanamid Co.

Columns. Column preparation is described in detail elsewhere
($\underline{12}$-$\underline{14}$). The polymers were coated from solution onto Chromosorb
G (AW-DMCS treated, 70/80 mesh) at approximately 10 wt-% loading.
For example, in the case of polystyrene, 2.7 g of polymer were
dissolved in 150 ml of benzene, 23 g of support were added, and
then the solvent was slowly evaporated by gently heating the
slurry(while constantly stirring). The coated support was then
dried in a vacuum oven (80°C) for 4 days and resieved before use
(60/80 mesh). The percent loading of polymer on support was
determined by calcination of 1 to 1.5 g of coating. A correction
was made for the loss of volatiles from the uncoated support. The
relative concentration of polymers in the blends was assumed to
be identical with that in the original solution prior to
deposition on the support. Columns were prepared from 48 mm
(internal diameter) copper tubing (typically 152 cm long) that
was plugged at each end by silanized glass wool . To provide
even packing, the column was constantly vibrated during filling.
Columns were conditioned under N_2 at temperatures above the glass
transition (usually T_g + 100°C) for a few hours before use.

Instrumentation. IGC measurements were carried out on a Hewlett-
Packard 5830A gas chromatograph equipped with a dual flame
ionization detector. The experimental set-up and procedure have
been described previously ($\underline{12}$-$\underline{14}$). Very small volumes (<0.01µL)
of the probe (together with the marker, methane) were injected
manually with a 10 µL Hamilton syringe in order to approach
infinite dilution conditions for the probe. Most of the probes

were characterized by symmetrical elution peaks and generally
exhibited little sample size dependence at low injection volumes,
low carrier gas flow rates (5 to 20 cm3/min), and moderate column
loading. Galin and Rupprecht (15) have shown that under these
conditions, the opposing influences of surface adsorption and gas
flow rate are nearly equivalent, so that the experimental $V_g°$
value is close to the bulk $V_g°$ value.
Data Reduction: Specific retention volumes, $V_g°(cm3/g)$, were
computed in a manner described elsewhere (12-14,16).

$$V_g^0 = t_N FJ/W_L \tag{1}$$

where t_N is the net retention time for the probe, F is the
carrier gas flow rate at 0°C and 1 atm (STP), J is a correction
factor for gas compressibility, and W_L is the weight of polymer
in the column.
    At temperatures above T_g, the magnitude of $V_g°$ is a measure
of the solubility of the probe in the stationary phase. From the
Flory-Huggins treatment of solution thermodynamics, one can
obtain the χ parameter, which is a measure of the residual free
energy of interaction between the probe and the polymer (17, 18).
    The relationship between χ and $V_g°$ (at T >T_g) is the
following:

$$\chi_{12} = \ln(273.16\, Rv_2/V_g^0 p_1^0 V_1) - (1 - V_1/V_2)\phi_2 - p_1^0(B_{11} - V_1)/RT \tag{2}$$

where v_2, V_2 and ϕ_2 refer to the specific volume, molar volume,
and volume fraction of the polymer, V_1 and $p_1°$ refer to the probe
molar volume and saturated vapor pressure respectively, R is the
gas constant, and T is the column temperature (K). B_{11} is the
second virial coefficient which is used to correct for vapor
phase non-ideality of the probe. Values of B_{11} were estimated
from corresponding equations of state (19, 20). Probe vapor
pressures were obtained from Dreisbach's compilation (21). Probe
densities were obtained from various sources, including the
compilations by Orwoll and Flory (22), Timmermans (23) and
International Critical Tables (24). For high molecular weight
polymer and infinite dilution of the probe, the second term of
Equation 2 [that is $(1-V_1/V_2)\phi_2$] approaches 1.
    It has been shown (3) using Scott's ternary solution
treatment (25) of the Flory-Huggins theory, that the overall
interaction parameter between the volatile probe (1) and the
binary stationary phase (2,3) is given by

$$\chi_{1(23)} = \ln([273.16R(w_2v_2 + w_3v_3)/V_g^0 p_1^0 V_1] - (1 - V_1/V_2)\phi_2$$

$$- (1 - V_1/V_3)\phi_3 - p_1^0(B_{11} - V_1)/RT \tag{3}$$

where w_2 and w_3 refer to the weight fractions of each polymer in
the blend.
    The volumetric data for the blends were determined by
assuming that the specific volume of the blend is the average of
the specific volumes of the parent homopolymers (26-29).

Results and Discussion

Polymer-polymer interaction parameters (χ_{23}') were calculated using the following expression:

$$X_{1(23)} = X_{12}\Phi_2 + X_{13}\Phi_3 - X_{23}'\Phi_2\Phi_3 \tag{4}$$

where 1 refers to the probe, 2 and 3 refer to the polymers in the stationary phase, ϕ_2 and ϕ_3 refer to the volume fraction of each of the polymers, and $\chi_{23}' = \chi_{23}V_1/V_2$, where V_1 and V_2 refer to the molar volume of the polymers. The value of χ_{23}' is thus normalized to the size of the probe molecule. A negative interaction parameter is required in order to ensure miscibility of two high molecular weight polymers.

Polymer-polymer interaction parameters are summarized for three systems:

1. blends of oligomeric polystyrene (PS$_L$) and poly(n-butyl methacrylate) (15 to 80 wt-% PS$_L$) at 140°C;
2. blends of polystyrene and poly(2,6-dimethyl-1,4-phenylene oxide) (25 to 85 wt-% PS) at 240°C; and
3. blends of semi-crystalline poly(vinylidene fluoride) and poly(methyl methacrylate) (25 to 90 wt-% PVF$_2$) at 200°C.

Tables I to III summarize the χ_{23}' values obtained with a variety of probes for each of these systems.

TABLE I. Polymer-Polymer Interaction Parameters (χ_{23}') for Various PS$_L$-PnBMA Blends at 140°C

Solute	Wt-% PS$_L$						
	15	27	30	35	40	58	80
n-octane	-0.11	0.10	0.42	-0.21	0.01	0.11	0.07
2,2,4-trimethylpentane	-0.25	0.09	0.47	-0.21	0.02	0.19	0.14
n-decane	-0.22	0.10	0.40	-0.20	-0.01	0.06	0.00
3,4,5-trimethylheptane	-0.21	0.12	0.43	-0.18	0.03	0.07	0.06
cyclohexane	-0.25	0.04	0.44	-0.24	-0.03	0.07	0.04
benzene	-0.17	0.11	0.47	-0.20	0.00	0.05	0.00
carbon tetrachloride	-0.22	0.05	0.45	-0.24	-0.02	0.09	0.08
chloroform	-0.25	0.09	0.41	-0.21	-0.03	0.09	0.01
2-pentanone	-0.30	0.08	0.40	-0.25	-0.06	0.02	-0.09
1-butanol	-0.41	0.06	0.35	-0.32	-0.08	-0.01	-0.03
n-butyl acetate	-0.25	0.08	0.43	-0.23	-0.08	0.04	0.08

TABLE II. Polymer-Polymer Interaction Parameters
(χ_{23}') for Various PS/PPO Blends at 240°C

Solute	Wt-% PS			
	25	50	75	85
n-octane	0.46	0.38	-0.52	-0.55
n-decane	0.38	0.53	-0.36	-0.40
3,4,5-trimethylheptane	1.32	0.86	-0.03	0.07
n-butylcyclohexane	0.62	0.60	-0.32	-0.31
cis-decalin	0.74	0.71	-0.19	-0.06
toluene	0.47	0.51	-0.19	-0.06
n-butylbenzene	0.54	0.46	-0.34	-0.34
chlorobenzene	0.48	0.49	-0.31	-0.21
acetophenone	0.40	0.49	-0.23	-0.05
cyclohexanol	0.66	0.58	-0.03	0.17

Source: Reprinted with permission from ref. 13.
Copyright 1985 Canadian Journal of Chemistry.

TABLE III. Polymer-Polymer Interaction Parameters
(χ_{23}') for Various PVF$_2$-PMMA Blends at 200°C

Solute	Wt-% PVF$_2$			
	25	50	75	90
acetophenone	0.55	-0.13	-0.51	-0.71
cyclohexanone	0.24	0.11	-0.33	-0.52
N,N-dimethylformamide	0.29	-0.20	-0.31	-0.45
cyclohexanol	0.03	-0.02	-0.46	-0.55
n-butylbenzene	0.12	0.01	-0.50	-0.59
o-dichlorobenzene	-0.01	-0.09	-0.50	-0.67
1-chlorooctane	0.06	0.08	-0.33	-0.60
1-chlorodecane	0.26	0.03	-0.47	-0.54

Source: Reprinted from ref. 14. Copyright 1982
American Chemical Society.

From these data, two general observations can be made.

1. As noted in previous chromatographic investigations of polymer-polymer miscibility ($\underline{3}$, $\underline{9}$, $\underline{10}$), some probe-to-probe variations are observed in each of these systems. The work of Al-Saigh and Munk ($\underline{7}$) and Pottiger ($\underline{30}$) indicates that this probe-to-probe variability is not intrinsic to the IGC technique, but is probably a limitation of the ability of the modified Flory-Huggins theory to account for all polymer-probe interactions in ternary solution systems (for example, inadequate expression for entropy of mixing which does not take into account non-random mixing of components). One might speculate that the probe-to-probe variation may indeed reflect true changes in interactions between the components of the stationary phases, due to the variations in force-fields at contact interfaces brought on by non-random partitioning of the probe molecules. The IGC technique may be unique in giving information on thermodynamic quantities as viewed from molecular, rather than bulk levels.

2. The χ_{23}' parameter is clearly dependent on the composition of the polymer blend. Examination of the tabulated data (Tables I to III) indicates that for each blend, all the probes yield similar trends. This composition dependence is illustrated graphically in Figures 1 to 3, where each point represents the average χ_{23}' value for all the probes investigated for each blend composition. (In the PS-PPO system, the probe 3,4,5-trimethylheptane exhibited large deviations and was therefore not considered in the averaging procedure.) This averaging procedure was employed in order to circumvent the variability in the χ_{23}' values and to facilitate illustration of the composition dependence.

IGC studies ($\underline{12\text{-}14}$) for each of these polymer blends reveal single, composition dependent T_g values (Figures 4-6), and in the case of PVF$_2$-PMMA blends, melting point depression is also observed (Figure 7). These are taken as indicators of polymer compatibility.

Blends of oligomeric polystyrene and poly(n-butyl methacrylate) are characterized by a large and unexpected variation of χ_{23}' as a function of blend composition (at 140°C). The large fluctuation in χ_{23}' between 20 and 40 wt-% PS$_L$ is difficult to explain. One of the referees has suggested that since the trend is the same for all the probes, a possible error in the measurement of some quantity common to all probes, such as the determination of the amount of polymer on the column, could explain these fluctuations. This remains to be confirmed. Since the measured values of χ_{23}' are generally positive, it appears that there are no strong attractive forces between these two polymers which would favor miscibility. However, because of the low molecular weight of the polystyrene, miscibility is permitted, even in the presence of positive χ_{23}' interaction parameters, due to favorable combinatorial entropy effects. Increasing the molecular weight of polystyrene leads to an immiscible system ($\underline{12}$).

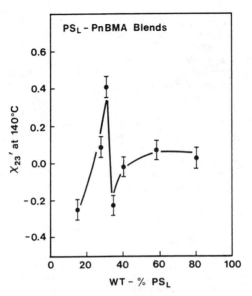

Figure 1.   Composition dependence of χ_{23}' in PS_L-PnBMA blends.
             (Reproduced from ref. 12. Copyright 1981 American
             Chemical Society.)

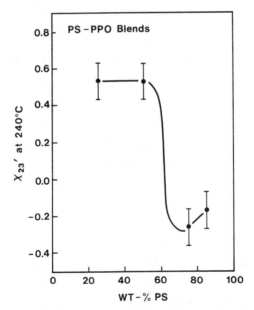

Figure 2.   Composition dependence of χ_{23}' in PS-PPO blends
             (Reproduced with permission from Ref. 13. Copyright
             1985 Canadian Journal of Chemistry.)

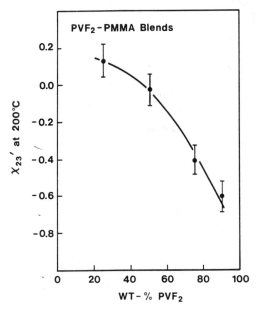

Figure 3. Composition dependence of χ_{23}' in PVF_2-PMMA blends.
(Reproduced from ref. 14. Copyright 1982 American
Chemical Society.)

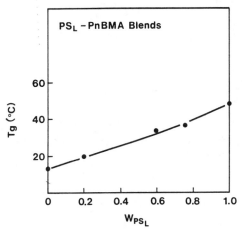

Figure 4. Composition dependence of T_g of PS_L-PnBMA blends.
(Reproduced from ref. 12. Copyright 1981 American
Chemical Society.)

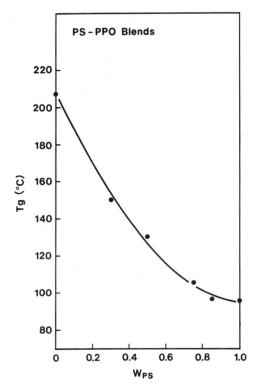

Figure 5.   Composition dependence of T_g of PS-PPO blends.
             (Reproduced with permission from Ref. 13. Copyright
             1985 Canadian Journal of Chemistry.)

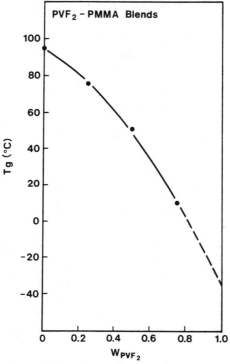

Figure 6. Composition dependence of T_g of PVF$_2$-PMMA blends. (Reproduced from ref. 14. Copyright 1982 American Chemical Society.)

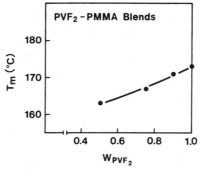

Figure 7. Composition dependence of T_m of PVF$_2$-PMMA blends. (Reproduced from ref. 14. Copyright 1982 American Chemical Society.)

Polystyrene-poly(2,6-dimethyl-1,4-phenylene oxide) blends with a high polystyrene content (>60 wt-% PS) are characterized by small negative interaction parameters (approximately -0.2) in the molten state. This is in accordance with the compatibility of PS-PPO blends. Small negative interaction parameters (< -0.1) have previously been reported for PS-PPO blends from melting point depression (31-33) and small-angle neutron scattering measurements (34). In addition, calorimetric studies have indicated a small negative enthalpy of mixing for this system at room temperature (35). In the present study, blends with low polystyrene content (<60 wt-% PS) are characterized by positive χ_{23}' values, indicating some microheterogeneity for these compositions. However, Fried and Su (36) have recently reported a negative interaction parameter for a 50 wt-% PS-PPO blend at 260°C from IGC measurements. Further work is required to identify specific reasons for this apparent difference.

Poly(vinylidene fluoride)-poly(methyl methacrylate) blends are characterized by a χ_{23}' parameter at 200°C which becomes more negative with increasing PVF_2 content, that is, from approximately 0.1 at 25 wt-% PVF_2 to -0.6 at 90 wt-% PVF_2. The negative χ_{23}' values at high PVF_2 content, indicating strong intermolecular interactions, are consistent with the melting point depression data of Nishi and Wang (37). A similar composition dependence has been reported by Wendorff (38) from small-angle X-ray scattering and melting-point depression studies of melt-blended mixtures of PVF_2 and PMMA. Small-angle neutron scattering experiments by Hadziioannou and Stein have also yielded negative interaction parameters for PVF_2-PMMA blends (39).

In summary, IGC is an experimentally attractive method for obtaining polymer-polymer interaction parameters in polymer blends at temperatures above T_m for a crystalline blend, and above T_g for an amorphous blend. This technique yields interaction parameters that are generally consistent with data obtained with other techniques such as vapor sorption, melting point depression, neutron scattering, and small-angle X-ray scattering (40). Advances in IGC of polymer blends will require increased experimental precision in order to improve the consistency of the data, as well as refinements of thermodynamic models to allow better interpretation of interactions in ternary solutions.

Acknowledgments

The experimental assistance of P. Degré, J. Richer and S. Fletcher is gratefully acknowledged.

Literature Cited

1. Braun, J. M.; Guillet, J. E. Adv. Polym. Sci. 1976, 21, 108.
2. Gray, D. G. Prog. Polym. Sci. 1977, 5 ,1.
3. Deshpande, D. D.; Patterson, D.; Schreiber, H. P.; Su, C. S. Macromolecules 1974, 7, 530.
4. Galin, M.; Rupprecht, M.C. Macromolecules 1979, 12, 506.

5. Klotz, S.; Schuster, R. H.; Cantow, H. J. Makromol. Chem. 1986, 187, 1491.
6. Su, C. S.; Patterson, D. Macromolecules, 1977, 10, 708.
7. Al-Saigh, Z.Y.; Munk, P. Macromolecules 1984, 17, 803.
8. Galin, M.; Maslinko, L. Eur. Polym. J. 1987, 23, 923.
9. Su, C.S.; Patterson, D.; Schreiber, H.P. J. Appl. Polym. Sci. 1976, 20, 1025.
10. Olabisi, O. Macromolecules 1975, 8, 316.
11. Ward, T.C.; Sheehy, D.P.; Riffle, J. S.; McGrath, J. E. Macromolecules 1981, 14, 1791.
12. DiPaola-Baranyi, G.; Degré, P. Macromolecules 1981, 14, 1456.
13. DiPaola-Baranyi, G; Richer, J.; Pest, Jr., W. M. Can. J. Chem. 1985, 63, 223.
14. DiPaola-Baranyi, G.; Fletcher, S. J.; Degré, P. Macromolecules 1982, 15, 885.
15. Galin, M.; Rupprecht, M.C. Polymer 1978, 19, 506.
16. Littlewood, A. B.; Phillips, C. S.G.; Price, D.T. J. Chem. Soc. 1955, 1480.
17. Patterson, D.; Tewari, Y. B.; Schreiber, H. P.; Guillet, J. E. Macromolecules 1971, 4, 356.
18. Flory, P. J. Principles of Polymer Chemistry; Cornell University Press: New York, 1953.
19. McGlashan, M. L.; Potter, D. J. B. Proc. R. Soc. London. Ser.A 1962, 267, 478.
20. Guggenheim, E. A. Wormald, C.J. J. Chem. Phys. 1965, 42, 3775.
21. Dreisbach, R. R. In Physical Properties of Chemical Compounds; Advances in Chemistry Series No. 15; American Chemical Society: Washington, DC, 1955; No.22, 1959; No. 29, 1961.
22. Orwoll, R. A.; Flory, P.J.; J. Am. Chem. Soc. 1967, 89, 6814.
23. Timmermans, J. In Physico-Chemical Constants of Pure Organic Compounds; Elsevier:New York, 1950; Vol. 1. 1965; Vol. 2.
24. International Critical Tables; McGraw Hill: New York, 1928; Vol.3.
25. Scott, R. L. J. Chem. Phys. 1949, 7, 279.
26. Hocker, H.; Blake, G.J.; Flory, P.J. Trans. Faraday Soc. 1971, 67, 2251.
27. Olabisi, O.; Simha, R. Macromolecules 1975, 8, 206.
28. Lewis, O.G. In Physical Constants of Linear Homopolymers; Springer-Verlag; New York, 1968.
29. Brandrup, J.; Immergut, E.H., Eds.; Polymer Handbook, 2nd ed.; Wiley: New York, 1975.
30. Pottiger, M.T., Ph.D. Thesis, University of Massachusetts, 1986.
31. Kwei, T.K.; Frisch, H. L. Macromolecules 1978, 11, 1267.
32. Shultz, A.R.; McCullough, C.R. J. Polym. Sci. Part A-2, 1972, 10, 307.
33. Berghmans, H.; Overbergh, N. J. Polym. Sci. Part A-2, 1977, 15, 1757.
34. Maconnachie, A.; Kambour, R.P.; White, D.M.; Rostami, S.; Walsh, D. J. Macromolecules 1984, 17, 2645.

35. Weeks, N. E.; Karasz, F. E.; MacKnight, W.J. J. Applied
     Phys. 1977, 48, 4068.
36. Fried, J. R.; Su, A.C. Polym. Mat. Sci. Eng. 1988, 58, 928.
37. Nishi, T.; Wang, T.T. Macromolecules 1975, 8, 909.
38. Wendorff, J. H., J. Polym. Sci., Polym. Lett. Ed. 1980, 18,
     439.
39. Hadziioannou, G.; Stein, R.S. Macromolecules 1984, 17, 567.
40. Riedl, B.; Prud'homme, R.E. Polym. Eng. Sci. 1984, 24, 1291.

RECEIVED September 29, 1988

Chapter 10

Inverse Gas Chromatography of Polymer Blends

Theory and Practice

Mohammad J. El-Hibri[1], Weizhuang Cheng[2], Paul Hattam, and Petr Munk

Department of Chemistry and Center for Polymer Research, University of Texas, Austin, TX 78712

With careful experimental design, inverse gas chromatography can be a viable method for the determination of the polymer-polymer interaction coefficient B_{23}. The variation of apparent B_{23} values with the probe is shown to be related to the chemical nature of the probe and not due solely to experimental error. A method is presented to allow the estimation of the 'true' B_{23} value. Experiments were performed on a 50/50 blend of poly(epichloro-hydrin)/poly(ϵ-caprolactone) at several temperatures. Polymer and blend solubility parameters were determined.

Probing polymer-polymer interactions in miscible blends is an experimentally difficult task. Most methods available for this purpose are elaborate and limited in their applicability. In recent years, research has shown that inverse gas chromatography (IGC) offers great promise for the study of polymer-polymer interactions. Conceptually, the technique involves the following: the elution behavior of volatile organic compounds (probes) is measured for one or more blend columns and compared with the retention behavior of two homopolymers studied under identical conditions. An excess retention can then be characterized and treated as a measure of polymer-polymer interaction strength. This polymer-polymer interaction is the cause of the miscibility phenomenon and is of practical interest.

Earlier attempts at using IGC to characterize polymer blends were unsuccessful. The polymer-polymer interaction parameters evaluated were found to vary with the probe used (1-5). For this reason, the use of IGC for the study of blends has been severely

[1]Current address: Amoco Performance Products, P.O. Box 409, Bound Brook, NJ 08805
[2]Current address: Department of Materials, Building Materials College, Shanghai, People's Republic of China

neglected. Given the importance of polymer interaction data and its unavailability through other methods, a thorough investigation of the technique was undertaken. A refined methodology for obtaining the experimental data of interest was implemented. This new methodology, is in part, the subject of this paper. The system poly(ϵ-caprolactone) /poly(epichlorohydrin) (PCL/PECH), a known compatible blend, was studied over the temperature range of 80 to 120°C. (For the comparison of data, results from an earlier work (6) on blends of composition PCL/PECH 25/75 and PCL/PECH 75/25 measured at 80°C have been included). Twenty-five probes, representing a number of chemical families, were used to examine the chemical contribution of the probe to the apparent value of the polymer-polymer interaction parameter derived from the IGC data.

Theory

The elution behavior of a probe on an IGC column is routinely described by the specific retention volume V_g , defined as

$$V_g = (V_r - V_o)/w \equiv V_n/w \tag{1}$$

where V_r is the probe elution volume, V_o is the void volume of the column, V_n is the net retention of the column, and w is the mass of the polymer. Combining the Flory-Huggins theory with standard chromatographic calculations, the probe-polymer interaction parameter X_{12} can be written as (7)

$$X_{12} = \ln (RTv_2/V_gV_1P_1^\bullet) - 1 + V_1/M_2v_2 - (B_{11} - V_1)P_1^\bullet/RT \tag{2}$$

In Equation 2, V_1 and v_2 are the probe molar volume and polymer specific volume, respectively; M_2 is the polymer molecular weight and R is the gas constant. $P_1^\bullet$ is the probe vapor pressure and B_{11} is its second virial coefficient in the gas phase. For work with high polymers, the third term of Equation 2 becomes negligible and may be omitted.

    Guillet and coworkers (8-10) have determined the solubility parameter of polymers from the probe-polymer interaction coefficients. They separated the interaction parameter into entropic and enthalpic contributions, such that $X_{12} = X_H + X_S$ to yield, in combination with Hildebrand's solution theory, the following expression:

$$X_{12} = V_1(\delta_1 - \delta_2)^2/RT + X_S \tag{3}$$

where δ_1 and δ_2 are the solvent and polymer solubility parameters, respectively. By expanding the expression in parentheses, they obtained the linear expression

$$\delta_1^2/RT - X_{12}/V_1 = (2\delta_2/RT)\delta_1 - (\delta_2^2/RT - X_S/V_1) \tag{4}$$

The experimental value of the left of Equation 4 was plotted against δ_1 . A value of δ_2 and an average value of X_S/V_1 were obtained by regression analysis.

    This method can be applied to blends, considering the blend as a single component and having a solubility parameter δ_{23}; the interaction coefficient being written as $X_{1(23)}$. In following the Guillet approach we did not separate X_{12} into entropic and enthalpic components. Instead, a more general term, $C_{1(23)}$, was introduced to represent a combination of interaction terms. By introducing the contact energy per unit volume, $B_{1(23)}$, a simpler form of the expression is obtained:

$$B_{1(23)} = RT X_{1(23)} / V_1 = (\delta_1 - \delta_{23})^2 + C_{1(23)} \qquad (5)$$

Rearrangement and expansion of Equation 5 yields the linear expression

$$\delta_1^2 - B_{1(23)} = 2\delta_{23}\delta_1 - \delta_{23}^2 - C_{1(23)} \qquad (6)$$

Thus, a plot of $\delta_1^2 - B_{1(23)}$ versus δ_1 yields $2\delta_{23}$ from the slope and an average value of $C_{1(23)}$ from the intercept.

When using IGC for the evaluation of the polymer-polymer interaction coefficient X_{23}, the free energy of mixing is routinely expressed by an extension of the Flory-Huggins expression (11) to a three component system (12):

$$\Delta G_{mix} = RT\left[n_1 \ln\phi_1 + n_2 \ln\phi_2 + n_3 \ln\phi_3 + n_1 \phi_2 X_{12} + n_1 \phi_3 X_{13} + n_2 \phi_3 X_{23}\right] \qquad (7)$$

where n_i, ϕ_i, and X_{ij} are the number of moles, volume fraction, and binary interaction parameter, respectively. An alternate parameter X_{23}', related to X_{23} as

$$X_{23}' = (V_1/V_2) X_{23} \qquad (8)$$

is conventionally used to describe the polymer-polymer interaction term as it removes the rather large value of the molar volume of the polymer, V_2. Routine thermodynamic calculations yield the expression for X_{23}' as

$$X_{23}' = (1/\phi_2\phi_3)\{\ln[V_{g,b}/(W_2 v_2 + W_3 v_3)] - \phi_2 \ln(V_{g,2}/v_2) - \phi_3 \ln(V_{g,3}/v_3)\} \qquad (9)$$

where the subscripts of V_g refer to the blend and to the homopolymers. W_2 and W_3 refer to the weight fractions of the two polymers in the blend. The interaction parameter may be given in terms of the contact energy per unit volume of the blend using the quantity B_{23}, in which X_{23}' is normalised with respect to the probe molar volume V_1:

$$B_{23} = RT X_{23}' / V_1 \qquad (10)$$

B_{23} is expected to be independent of the nature of the probe.

Improvements in the IGC Method

It was recognized that the levels of precision and reproducibility
adequate in IGC studies of homopolymers were inadequate for a
successful study of blend systems. A column-to-column
reproducibility of 1% was deemed necessary for this purpose. This
is because the quantity of interest in the case of blends is the
difference between the retentions of the blend column and the
homopolymer columns, which is usually less than 10% of the observed
retention for any of the individual columns. Thus, a number of
experimental and data analysis improvements has been introduced to
the technique, which have boosted the reproducibility of the data
considerably.

Experimental Modifications. Perhaps the most significant change
introduced is the mode of coating the polymer onto the inert
packing. Traditionally, the polymer sample is deposited onto the
support in solution, using slow solvent evaporation. This method
has the disadvantage of preventing precise determination of the
polymer mass due to losses of polymer on the walls of the
preparation vessel. Calcination and Soxhlet extraction, performed
for subsequent mass determination, have been shown to be major
causes of error (13,14). We used a new coating technique, (partial
soaking method), which consists of the following steps. The
polymer is first dissolved in a suitable solvent. The support is
then piled on a watch glass and a portion of the solution added to
the top of the support pile. Care is exercised so that the
solution does not come in contact with the watch glass. The
support is thoroughly mixed and the process repeated until all the
solution has been used (including several rinsings of the solution
flask). Consequently the exact mass of the polymer coated onto the
support is known. The procedure has been described in detail
elsewhere (15). Two other experimental aspects were modified. The
precision in measuring the carrier gas flow rate was enhanced by a
new soap-bubble flow meter design (16). Also, the resolution of
the detection of the elution data was improved by implementing a
custom-configured computer-based data handling system. In this
scheme, an HP-3478A digital voltmeter was interfaced with a
microcomputer using an IEEE-488 interface board (National
Instruments) and the detector output monitored. This configuration
allows elution data to be measured with a signal-to-noise ratio of
5×10^4 in the detector output reading. Elution times are measured
with a precision of ± 0.1 s.

Data Analysis. A tacit assumption that the support material
contributes little or no retention to the observed retention by the
polymeric coating is usually made in the IGC literature. In a
published work (17), and from a large body of recently gathered
data, it has been confirmed that retention by the so-called inert
support may actually account for up to 10% of the observed
retention of the column. Furthermore, the support retention was
found to be a function of the amount of probe injected, especially
for strongly polar probes. It became clear that this factor alone
could undermine the blend analysis if it were not handled properly.

A procedure was developed in which the retention by the polymer was obtained by subtracting the retention of the support from the observed retention of a given column V_g^{obs}. According to this treatment, the specific retention volume is given by

$$V_g = V_g^{obs} - V_n^{sup}/w \qquad (11)$$

where V_n^{sup} is the retention volume of the support, as obtained from an independent experiment on an uncoated support column under identical conditions. The fundamental assumption made in Equation 11, the additivity of the support and polymer retentions, is strongly supported by our experimental data.

The concentration dependence of the support retention for the various probes followed the relation

$$\ln V_n^{sup} = \alpha + \beta \ln A \qquad (12)$$

where A, the peak area, was used as a measure of the amount of probe injected. α and β are functions of temperature. The temperature dependence of α and β followed a dependence of the Arrhenius type. In the case of alkanes and other non-polar probes, V_n^{sup} was essentially independent of probe concentration; that is β was small or zero. This behavior was interpreted as a possible result of retention of these probes by the polymer polydimethylsiloxane, which can be formed on the support surface during its treatment with dimethylchloro-silane (DMCS). Polar probes retention is strongly dependent on the probe concentration ($\beta = -0.2$ to -0.5). This behavior was interpreted as interaction of polar probes with the few polar groups on the surface of the Chromosorb that were not removed during DMCS treatment.

The procedure for correcting the observed retention data involved the following steps. First, the area of the peak from a polymer-coated column was inserted into Equation 8 to determine the support retention corresponding to that particular area. Then the computed V_n^{sup} value was subtracted according to Equation 11. This procedure was found to yield high reproducibility for V_g data, which was unattainable otherwise.

Another correction in the retention volume was made to account for retention of the methane marker on the column. The retention volume of methane, V_n^m, was used for computing a corrected column void volume, V_o^C:

$$V_o^C = V_r^m - V_n^m \qquad (13)$$

where V_r^m is the marker elution volume. The quantity V_n^m was estimated by an iterative extrapolation of the retention data for normal alkanes correlated against alkane carbon number, n (<u>18</u>). A linear relationship between the logarithm of retention volume and the alkane number has been known for a long time (<u>19</u>), and so far, all IGC data have followed it. The marker correction is not as important as that for the support, but it is easy to perform and does improve the quality of the data, particularily for weakly retained probes.

Results and Discussion

Polymer-Polymer Interaction Parameter. Table I lists the probes
used in this work and the numbers that correspond to those shown in
the Figures.

Table I.   List of Compounds Used and Their Numbers
           Corresponding to Points in the Figures

No.	Probe	No.	Probe
3	Pentane	16	Acetone
4	Hexane	17	Butanone
5	Heptane	18	Methyl Acetate
6	Octane	19	Ethyl Acetate
7	Nonane	20	Propyl Acetate
8	Cyclohexane	21	Butyl Acetate
9	Benzene	22	1,1-Dichloroethane
10	Toluene	23	1,2-Dichloroethane
11	Methylene Chloride	24	1,1,1-Trichloroethane
12	Chloroform	25	Butyl Chloride
13	Carbon Tetrachloride	26	Pentyl Chloride
14	Tetrahydrofuran	27	Cyclohexene
15	Dioxane		

The polymer-polymer interaction parameters χ_{23}' and B_{23} are
presented in Table II for three blend compositions measured at 80°
C; included in Table II are the probe-polymer interaction
coefficients χ_{12} and χ_{13} for the homopolymers. For the PCL/PECH
blends, the measured values of interaction parameters for a given
probe do not vary with blend composition. The small differences
(<0.05) can be attributed to experimental error. The values differ
for different probes, though to a lesser extent than the values
reported in the literature. The general behavior is that, with
non-solvents or poor solvents, the interaction parameters have
relatively high values, whereas the polar probes give much lower
values. The temperature dependence of the interaction parameters
for the PCL/PECH 50/50 blend is presented in Table III over the
temperature range 80 to 120°C. Over this temperature range there
is, within our experimental error, no change in the interaction
parameters. However, the general trend between polar and non-polar
probes is retained.

The differences among B_{23} values for different probes are
interpreted as reflecting the inadequacy of the underlying
expression for ΔG_{mix}, see Equation 7. A more general expression
for ΔG_{mix} has been shown to predict that the apparent B_{23} value
calculated from Equations 9 and 10 depends on all possible
interactions in the ternary system probe-polymer-polymer and not
only on the polymer-polymer interaction (15). This problem has
been approached by examining the dependence of B_{23} on the
Hildebrand solubility parameter δ_1 of the probe (calculated from
the Antoine coefficients). Examples of this dependence are given
in Figures 1 and 2 for the blend PCL/PECH 50/50 at 80°C and at 120°
C, respectively. The curves tend to reach a minimum at a value of
the probe solubility parameter of approximately 9 $(cal/mL)^{\frac{1}{2}}$.

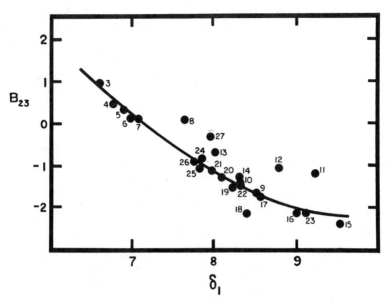

Figure 1. Dependence of the polymer-polymer interaction coefficient B_{23} (cal/mL) on the solubility parameter of the probe δ_1 (cal/mL)$^{\frac{1}{2}}$ for PCL/PECH 50/50 blend at 80°C.

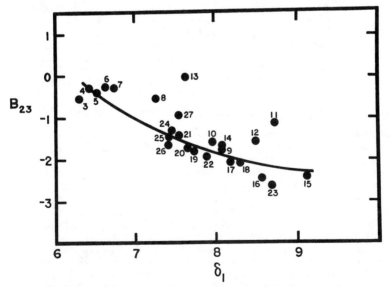

Figure 2. Dependence of the polymer-polymer interaction coefficient B_{23} (cal/mL) on the solubility parameter of the probe δ_1 (cal/mL)$^{\frac{1}{2}}$ for PCL/PECH 50/50 blend at 120°C.

Table II. Polymer-Polymer Interaction Parameters X_{23}' and B_{23}
for PCL/PECH Blends at Three Compositions and the
Interaction Coefficients for the Homopolymers
Determined at 80°C

	PCL/PECH X_{23}'			PCL/PECH B_{23} (cal/mL)			PCL X_{12}	PECH X_{13}
	Volume Fraction of PCL							
	.25	.50	.75	.25	.50	.75	1.0	0.0
Probe No.								
3	.10	.08	.00	0.6	0.4	0.0	1.19	1.74
4	.11	.05	.06	0.6	0.3	0.3	1.24	1.82
5	.11	.05	.05	0.5	0.2	0.2	1.31	1.92
6	.12	.05	.06	0.5	0.2	0.2	1.39	2.04
7	.12	.05	.06	0.4	0.2	0.2	1.47	2.16
8	.07	.02	.00	0.4	0.1	0.0	0.94	1.32
9	-.19	-.22	-.20	-1.4	-1.6	-1.4	0.04	0.23
10	-.19	-.22	-.21	-1.2	-1.3	-1.3	0.06	0.03
11	-.09	-.12	-.06	-0.8	-1.0	-0.5	-0.51	0.22
12	-.19	-.19	-.19	-1.9	-1.9	-1.9	-0.31	0.20
13	-.02	-.08	-.05	-0.1	-0.5	-0.3	0.24	0.73
14	-.12	-.16	-.16	-0.9	-1.3	-1.3	0.12	0.00
15	-.25	-.29	-.27	-2.0	-2.3	-2.2	0.11	0.00
16	-.25	-.26	-.24	-2.0	-2.3	-2.1	0.48	0.27
17	-.20	-.24	-.24	-1.4	-1.7	-1.7	0.35	0.18
18	-.16	-.25	-.23	-1.3	-2.0	-1.9	0.39	0.36
19	-.18	-.23	-.22	-1.2	-1.5	-1.5	0.35	0.33
20	-.18	-.22	-.23	-1.0	-1.3	-1.3	0.32	0.32
21	-.19	-.21	-.23	-0.5	-1.0	-1.1	0.30	0.34
22	-.17	-.18	-.15	-1.3	-1.4	-1.2	-0.06	0.36
23	-.24	-.24	-.21	-1.9	-2.0	-1.7	-0.18	0.19
24	-.08	-.12	-.09	-0.5	-0.8	-0.6	0.06	0.49
25	-.13	-.16	-.13	-0.8	-1.0	-0.8	0.35	0.65
26	-.11	-.15	-.13	-0.6	-0.8	-0.7	0.35	0.70
27	.00	-.04	-.04	0.0	-0.3	-0.2	0.61	0.92

As discussed below this value is close to the blend solubility
parameter determined from the method initiated by Guillet (19),
that is according to Equation 6.

Two things are evident from the dependencies. One, the
solubility parameter allows a reasonable correlation with B_{23}.
Two, the correlation is far from perfect; a number of probes
deviates from the line outside of the experimental error. (In the
worst case the error in B_{23} is ±0.25 cal/mL.) While six points
deviate considerably from the line, 19 points lie within
experimental error. Furthermore, most probes that deviate
significantly from the line have chemical groups that can interact
strongly with the chemical groups present in the blend. Thus this
deviation appears to be a result of specific interactions rather
than experimental scatter.

It is important to determine how the true interaction
parameter B_{23} is related to the values presented in Figure 1 and
Figure 2. We are surmising that the true value would be shown by a

hypothetical probe incapable of specific interactions with the two polymers and having the same solubility parameter as the blend. In other words, the value should be interpolated on the solid line of Figures 1 and 2 (this line presumably represents non-interacting probes) for a value of δ_1 equal to the solubility parameter of the blend δ_{23}.

Table III. Polymer-Polymer Interaction Parameters for 50/50 PCL/PECH Blend at Five Temperatures ($^\circ$C)

Probe	X_{23} 80	90	100	110	120	B_{23} (cal/mL) 80	90	100	110	120
3	.18	-.02	.12	-.07	-.10	1.0	-0.1	0.6	-0.4	-0.6
4	.10	.03	.07	.02	-.05	0.5	0.2	0.4	0.1	-0.3
5	.08	.00	.01	-.03	-.08	0.3	0.0	0.0	-0.2	-0.4
6	.04	.03	.02	.00	-.05	0.2	0.1	0.1	-0.0	-0.2
7	.04	.05	.01	.00	-.07	0.1	0.2	0.0	0.0	-0.3
8	.01	-.01	-.01	.01	-.09	0.1	-0.1	-0.1	0.1	-0.5
9	-.23	-.24	-.20	-.21	.24	-1.7	-1.8	-1.5	-1.6	-1.8
10	-.17	-.16	-1.7	-.16	-.21	-0.9	-0.9	-0.9	-0.9	-1.2
11	-.11	-.21	-.15	-.06	-.11	-1.2	-2.2	-1.5	-0.7	-1.2
12	-.13	-.14	-.12	-.09	-.18	-1.1	-1.2	-1.0	-0.8	-1.6
13	-.10	-.06	-.05	-.01	-.01	-0.7	-0.4	-0.3	-0.0	-0.1
14	-.16	-.18	-.17	-.16	-.21	-1.3	-1.4	-1.4	-1.3	-1.7
15	-.29	-.29	-.28	-.27	-.29	-2.4	-2.4	-2.4	-2.3	-2.5
16	-.25	-.28	-.17	-.11	-.27	-2.1	-2.4	-1.5	-1.0	-2.5
17	-.24	-.25	-.24	-.23	-.27	-1.7	-1.8	-1.8	-1.7	-2.1
18	-.26	-.28	-.22	-.22	-.24	-2.2	-2.3	-1.8	-1.8	-2.1
19	-.23	-.24	-.24	-.24	-.26	-1.5	-1.6	-1.6	-1.6	-1.8
20	-.23	-.23	-.23	-.22	-.30	-1.3	-1.3	-1.3	-1.3	-1.8
21	-.22	-.22	-.23	-.23	-.27	-1.1	-1.1	-1.2	-1.2	-1.4
22	-.25	-.25	-.25	-.19	-.31	-2.1	-2.1	-2.1	-1.7	-2.7
23	-.23	-.24	-.23	-.21	-.24	-1.4	-1.5	-1.4	-1.4	-1.6
24	-.19	-.20	-.19	-.17	-.25	-1.5	-1.5	-1.5	-1.4	-2.0
25	-.13	-.13	-.14	-.14	-.19	-0.8	-0.9	-0.9	-0.9	-1.3
26	-.17	-.15	-.17	-.17	-.22	-1.1	-0.9	-1.1	-1.1	-1.5
27	-.05	-.06	-.08	-.05	-.14	-0.3	-0.4	-0.5	-0.4	-1.0

The solubility parameters of the homopolymers and those of the blends have been determined using the method of Guillet and coworkers (8-10) as modified in Equation 6 above. The values of δ_{23} and $C_{1(23)}$ are presented in Tables IV and V. (A more detailed discussion of the Guillet method is presented in the next section). Using these values of the polymer solubility parameters, the polymer-polymer interaction parameter B_{23} has been determined. Since the value of δ_{23} is close to the value of δ_1 at the minimum of the curve B_{23} versus δ_1, only negligible error is caused by any erroneous estimate of δ_{23}. The data are presented in Table VI.

Though it is expected that the B_{23} value is dependent on temperature, this was not the case for our blend in the range of temperatures used (80 to 120°C); nor is any change observed with composition. Overall the value of B_{23} was -2.1 cal/mL, which seems to be a reasonable value for such a strongly interacting blend.

Table IV.   Polymer and Blend Solubility Parameters $(cal/mL)^{\frac{1}{2}}$

	PCL	PCL/PECH 25/75 Blend Average		PCL/PECH 50/50 Blend Average		PCL/PECH 75/25 Blend Average		PECH
T(°C)								
80	9.31	9.47	9.55	9.34	9.47	9.32	9.39	9.63
90	9.17	-	-	9.20	9.32	-	-	9.48
100	9.01	-	-	9.05	9.15	-	-	9.10
110	8.87	-	-	8.81	8.99	-	-	9.10
120	8.72	-	-	8.62	8.87	-	-	9.03

Table V.   Combinatory Interaction Parameter $C_{1(23)}$ (cal/mL)

	PCL	PCL/PECH 25/75	PCL/PECH 50/50	PCL/PECH 75/25	PECH
T(°C)					
80	0.07	1.09	1.12	0.91	0.95
90	0.14	-	1.21	-	1.00
100	0.28	-	1.24	-	1.22
110	0.32	-	1.75	-	1.40
120	0.39	-	1.52	-	1.17

Table VI.   Values of the Blend Interaction Parameter B_{23} (cal/mL)

	PCL/PECH 25/75	PCL/PECH 50/50	PCL/PECH 75/25
T(°C)			
80	-2.05	-2.08	-2.20
90	-	-2.24	-
100	-	-2.08	-
110	-	-2.08	-
120	-	-2.16	-

Polymer Solubility Parameters. In Table IV, the values of δ_{23} are shown (a) evaluated from the Guillet plot and (b) calculated as the volume fraction weighted average of the values for homopolymers. Since the solubility parameter is a measure of the cohesive forces present in the material it would be reasonable to assume that the average value would be lower than the value for the blend. However, the reverse was observed. We have therefore decided that the Guillet method deserves closer examination. To do this the equation of the straight line (Equation 6) was recast to obtain the Guillet smoothed dependence in the original coordinates $B_{1(23)}$ and δ_1. (It is a quadratic). Figures 3 and 4 show the Guillet straight line for the 50/50 blend at 80°C and 120°C and Figures 5 and 6 show the actual data; the line is the quadratic determined from the Guillet approach. There is considerable scatter of points: the apparently good correlation of the Guillet plot is actually quite deceptive. In the $B_{1(23)}$ versus δ_1 dependence, many points deviate from the line much more than our experimental error (which was estimated to be ± 0.25 cal/mL). We consider these deviations to be a result of non-dispersive

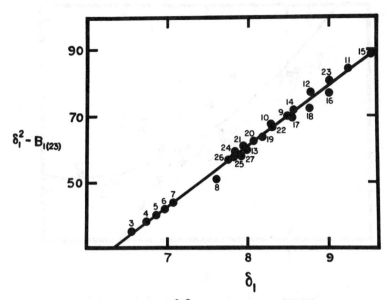

Figure 3. Dependence of $\delta_1^2 - B_{1(23)}$ on the solubility parameter of the probe δ_1 used for determining blend solubility parameter δ_{23} of PCL/PECH 50/50 blend at 80°C.

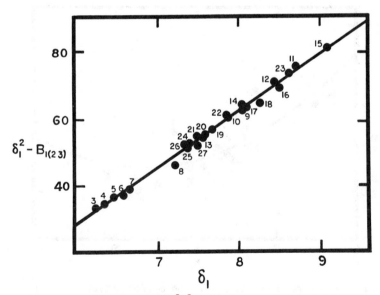

Figure 4. Dependence of $\delta_1^2 - B_{1(23)}$ on the solubility parameter of the probe δ_1 used for determining blend solubility parameter δ_{23} of PCL/PECH 50/50 blend at 120°C.

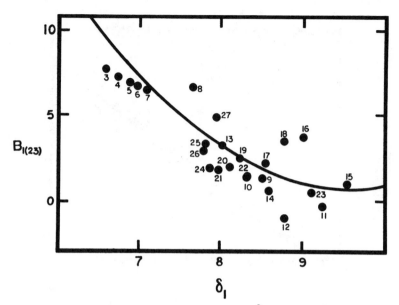

Figure 5. Dependence of $B_{1(23)}$ on δ_1 for PCL/PECH blend at 80°C. Solid line is the transform of the correlation line in Figure 3.

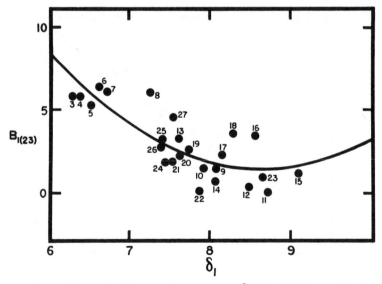

Figure 6. Dependence of $B_{1(23)}$ on δ_1 for PCL/PECH blend at 120°C. Solid line is the transform of the correlation line in Figure 4.

interactions of groups in the probe and in the polymer. Only dispersive interactions were considered in the Hildebrand theory of solutions that formed the basis for Equation 6. We have concluded that the exact position of the Guillet straight line reflects the selection of the probes and that the resulting value of δ_2 or δ_{23} must be considered only as an approximation. Nevertheless the values of δ_{23} are comparable to those obtained by other methods (20).

Conclusions

The following conclusions were drawn from the experiments conducted in this work.

1. IGC data of sufficient accuracy may be obtained only if certain experimental and acquisition/analysis techniques are followed.

2. With refinements in the technique, the method of IGC can be used to determine the polymer-polymer interactions in polymer blends.

3. It is postulated that a hypothetical probe, that has the same solubility parameter (cohesive energy) as the blend and does not exhibit any specific interactions with the blend components, yields the true polymer-polymer interaction coefficient.

4. The parameter of the cohesive energy, B_{23}, for the polycaprolactone-polyepichlorohydrin blend was found to be   -2.1 cal/mL. Within experimental error, B_{23} was independent of the composition of the blend and of the temperature in the range 80 to 120 $^\circ$C.

5. Specific interactions between the probe and polymer are not entirely accounted for in current theory, causing a deviation of experimental data from that predicted by the theory.

6. Determination of the polymer solubility parameter using the Guillet approach yields deceptively linear dependences compared to the scatter inherent in the experimental data. While the technique is more than adequate, compared with values obtained from more time consuming classical methods, one should be aware of the limitations of generating the linear dependence.

Acknowledgment

The authors are grateful for the financial support of the National Science Foundation through grant DMR-8414575.

Literature Cited

1. Olabisi, O. Macromolecules 1975, 8, 316.
2. Robard, A. and Patterson, D. Macromolecules 1977, 10, 1021.
3. Su, C. S.; Patterson, D. and Schreiber, H. P. J. Appl. Polym. Sci. 1976, 20, 1025.
4. Walsh, D. and McKeon, G. J. Polymer 1980, 21, 1335.
5. Su, A. C. and Fried, J. R. J. Polym. Sci., Pol. Lett. Ed. 1986, 24, 343.
6. El-Hibri, M. J.; Cheng, W. and Munk, P. Macromolecules 1988 (in Press).

7. Smidsrod, O. and Guillet, J. E. Macromolecules 1969, 2, 272.
8. DiPaola-Baranyi, G. and J. E. Guillet, Macromolecules 1978, 11, 228.
9. Ito, K. and Guillet, J. E. Macromolecules 1979, 12, 1163.
10. Price, G. J.; Guillet, J. E. and Purnell, J. H. J. Chromatog. 1986, 369, 273.
11. Flory, P. J. Discuss. Faraday Soc. 1970, 7, 49.
12. Scott, R. L. J. Chem. Phys. 1949, 17, 268.
13. Laub, R. J.; Purnell, J. H.; Williams, P. S.; Harbison, M. W. P. and Martire, D. E. J. Chromatog. 1978, 155, 233.
14. Ashworth, A. J.; Chien, C. F.; Furio, D. L.; Hooker, D. M.; Kopecni, M. M.; Laub, R. J. and Price, G. J. Macromolecules 1984, 17, 1090.
15. Al-Saigh, Z. Y. and Munk, P. Macromolecules 1984, 17, 803.
16. Card, T. W.; Al-Saigh, Z. Y. and Munk, P. J. Chromatog. 1984, 301, 261.
17. Card, T. W.; Al-Saigh, Z. Y. and Munk, P. Macromolecules 1985, 18, 1030.
18. El-Hibri, M. J. and Munk, P. Macromolecules 1988, 21, 264.
19. Guillet, J. E. J. Macromol. Sci. Chem. 1970, A4, 1669.
20. Guillet, J. E. and Price, G. J. J. Soln. Chem. 1987, 16, 605.

RECEIVED November 2, 1988

Chapter 11

Estimation of Free Energy of Polymer Blends

S. Klotz, H. Gräter, and H.-J. Cantow

Institut für Makromolekulare Chemie der Albert-Ludwigs-Universität
Freiburg, Hermann-Staudinger-Haus, Stefan-Meier-Strasse 31, D–7800
Freiburg, Federal Republic of Germany

Blends of polystyrene/poly(2,6-dimethyl-1,4-
phenylene oxide) and polystyrene/poly(vinyl
methyl ether) were investigated by IGC over
wide composition and temperature ranges.
Flory-Huggins free energy parameters were obtained and are discussed as the criterion for
thermodynamic miscibility. From the temperature variation of the free energy parameter,
phase diagrams for both blends were obtained.
IGC was shown to give a correct thermodynamic
interpretation of molten polymeric mixtures.

With the growing interest in polymer blends, a variety of
sophisticated experiments are being used to determine
polymer compatibility (1,2). Only a few techniques can
give quantitative information about the change in free
energy when mixing two polymers. Especially small angle
neutron scattering (SANS) experiments with mixtures of
deuterated and undeuterated polymers have been used to
measure the thermodynamic state of blends. In solution
thermodynamics, one standard procedure is to measure the
sorption of a low molecular weight solvent into a solid
polymer. The amount of solvent present in the polymer and
the solvent vapour pressure are determined by the chemical potential of the system. Since the vapour pressure oᴸ

0097–6156/89/0391–0135$06.00/0
© 1989 American Chemical Society

a polymer is low, there is no way to get direct information about the change in the free energy in polymeric
blends with this technique. This problem may be overcome
by using inverse gas chromatography (IGC), where a volatile low molecular weight probe is used.

Considering a ternary system of two polymers, 2 and
3, and a solvent, 1, the Flory-Huggins free energy parameter $\chi_{1(23)}$ in the single liquid approximation is gi'en
by ($\underline{3}$)

$$\chi_{1(23)} = \phi_2\chi_{12} + \phi_2\chi_{13} - \phi_2\phi_3\chi_{23}' \tag{1}$$

χ_{1i} and $\chi_{1(23)}$ represent the free energy parameter of the
binary systems solvent 1/polymer i (i=2 and 3), and the
quasi-binary system solvent 1/blend 23. Consequently,
χ_{23}' describes the polymer 2/polymer 3 interaction
energy, which cannot be directly measured. Yet, $\chi_{1(23)}$,
χ_{12}, and χ_{13} are experimentally accessible quantities, and
χ_{23}' may be calculated from Equation 1.

Unfortunately it is often observed that in a ternary
solution, the polymer 2/polymer 3 free energy parameter
χ_{23}' is influenced by the solvent used ($\underline{4}$-$\underline{6}$). Another
problem arises with the solvent/polymer χ parameters,
which are usually one order of magnitude larger than
χ_{23}'. Thus, great accuracy is needed to get correct
information about χ_{23}'. A third disadvantage is that the
three independent measurements to get χ_{23}' are time consuming. With these problems in mind, IGC was used to
investigate   the   two   polymer   pairs   polysty-
rene(PS)/poly(vinyl   methyl   ether)(PVME)   and   polysty-
rene(PS)/poly(2,6-dimethyl-1,4-phenylene   oxide)(PPE).
Both blends are known to be compatible at ambient temper-
ature ($\underline{7}$,$\underline{8}$). PS/PVME shows phase separation at high temperatures (Lower Critical Solution Temperature, LCST)
($\underline{9}$), and LCST-behavior was predicted for PS/PPE ($\underline{10}$).

Experimental Materials and Methods

Materials(PS/PVME). Atactic PS of weight-average molecular weight M_w=17,500 and polydispersity index M_w/M_n=1.06 was purchased from Pressure Chemical. PVME was obtained from Aldrich-Chemie, Steinheim, FRG, and was fractionated to give a sample of M_w=70,000 and M_w/M_n=1.6. The molecular weights were determined by gel permeation chromatography, light scattering, and osmometry. Reagent grade acetone, ethyl acetate, cyclohexane, n-octane, and ethyl benzene from Fluka and Roth were used without further purification.

Materials(PS/PPE). Atactic PS of molecular weight M_w=50,000 and M_w/M_n=1.05 was prepared in the laboratory PPE was purchased from Aldrich-Chemie, Steinheim, FRG, with M_w=46,000 and M_w/M_n=2.4. Reagent grade ethylbenzene, toluene, benzene, cyclohexane, and n-octane from Fluka and Roth were used without further purification.

Column Preparation. All stationary phases were coated onto Chromosorb W, HP (Supelco, mesh size 80 to 100, silane treated) by dissolution in toluene, stirring for 48 h, and slow evaporation of the solvent at 50°C under dry nitrogen. The coated support was dried in a vacuum oven at 100°C for 72 h and packed by a gentle tapping procedure into a stainless steel column (inner diameter 2 mm, length 1.8 m), the end of which was loosely plugged with glass wool. The tubing was conditioned for 3 days at 120°C under dry nitrogen until the weight of the column remained constant. After each experiment, the column loading was determined by calcination of the coated support material. Within experimental error, no weight loss was detected during a set of measurements.

Instrumentation. Measurements were carried out on a Perkin-Elmer Sigma 3 dual-column gas chromatograph, equipped with a thermal conductivity detector. Column temperature was controlled within +/-0.1°C. Probes were introduced by manual injection or with a heated sampling valve. The pressure at the inlet and outlet of the column were measured by a high performance pressure gauge (Wika, FRG). Dry, purified nitrogen was used as the carrier gas. A large range of carrier gas flow rates (1 to 20mL·min^{-1})

was measured within an error limit of 0.1% using calibrated soap film flow meters.

Measurement Procedure. IGC measurements were started
after the thermal and flow equilibrium in the column were
stable (2 to 3 h). To facilitate rapid vaporization of
the probe (0.01 µL), the injector temperature was kept
30°C above the boiling point of the probe. Measurements
were made at five carrier gas flow rates. The retention
volumes of six injections for each probe and twenty
injections of the marker (H_2) at a given flow rate were
averaged. The values obtained were extrapolated to zero
flow rate to eliminate the flow rate dependence of the
retention data. The net retention time (t_R) is defined as
the time difference between the first statistical moment
of the solvent peak and that of the marker gas. Thus, t_R
was calculated by an on-line computer statistical peak
analysis rather than the retention time at the peak maximum ($t_{R,max}$). This eliminated inaccuracies arising from
slight peak asymmetry, which occurs even for inert and
well-coated supports. The specific retention volumes
(V_g^0) derived from t_R and $t_{R,max}$ differed by as much as
5% for small retention times and slightly skewed peaks
(11,12).
Data Treatment. Reduced specific retention volumes (V_g)
were calculated from the expression (13)

$$V_g = [(273 t_R * F)/(T_R * m_2)] * (3/2) * [(p_i/p_0)^2 - 1]/[(p_i/p_0)^3 - 1] *$$

$$(1 - p_w/p_a) \qquad (2)$$

where t_R is the net retention time, m_2 is the mass of the
polymer in the column, F is the carrier gas flow rate at
room temperature and atmospheric pressure (p_a), T_R is the
room temperature, p_i and p_0 are the inlet and outlet
pressures, and p_w is the water vapour pressure at room
temperature. Values of V_g were measured at five carrier
gas flow rates and extrapolated to zero flow rate to
obtain V_g^0. The weight fraction activity coefficient of

the probe, Ω_1^∞, at the limit of zero concentration, is related to the specific volume as follows ($\underline{14}$):

$$\Omega_1^\infty = (273R/V_g^0 p_1^0 M_1) \exp(-p_1^0(B_{11}-V_1)/RT) \tag{3}$$

where M_1 represents the molecular weight of the probe, p_1^0 is the saturated vapour pressure, V_1 is the molar volume, and B_{11} is the second virial coefficient of the probe vapour. The activity coefficient Ω_1^∞ of the probe can be related to the reduced free energy parameter χ through conventional polymer solution theory, if the limiting case of polymer volume fraction $\phi \rightarrow 1$ is approached. For a homopolymer,

$$\chi_{1i} = \ln\Omega_1^\infty - (1-1/r_i) + \ln(v_i^{sp}/v_1^{sp}) \tag{4}$$

where r_i represents the mean number of segments per molecule i and v_1^{sp} and v_i^{sp} respresent the specific volumes of the probe and the polymer, respectively. Based on the single liquid approximation for a mixed stationary phase ($\underline{3}$)

$$\ln\Omega_1^\infty = \ln(v_1^{sp}/(v_2^{sp}w_2+v_3^{sp}w_3))+1$$

$$+\phi_2\chi_{12}+\phi_3\chi_{13}+\phi_2\phi_3\chi_{23}' \tag{5}$$

$\chi_{23}'=\chi_{23}V_1/V_2$, where V_1 and V_2 are the molar volumes of the probe and component 2. Combining Equations 3, 4, and 5 yields

$$\chi_{23}' = (\phi_1\phi_2)^{-1}*\ln(V_{g,23}^0/(v_2^{sp}w_2+v_3^{sp}w_3))$$

$$-\phi_3^{-1}*\ln(V_{g,2}^0/v_2^{sp}) -\phi_2^{-1}*\ln(V_{g,3}^0/v_3^{sp}) \tag{6}$$

$V_{g,2}{}^0$ and $V_{g,3}{}^0$ are the specific retention volumes of the probe in the molten homopolymers, and $V_{g,23}{}^0$ the specific retention volume of the probe in the blend. The physical data of the probes were taken from various sources (15,16). The densities of PS and PPE at elevated temperatures were obtained from Höcker et al. (17) and Hoehn et al. (18), respectively. The density and the thermal expansion coefficient of PVME were taken from dilatometric measurements (Klotz, S., University of Freiburg, unpublished data).

Results and Discussion

It is known that the column retention behavior of a probe depends on bulk absorption and surface adsorption (19). When the coated polymer film is thin, surface adsorption phenomena are pronounced. To minimize these effects, a series of different polymer loadings (3.85 wt-%, 6.67 wt-%, 8.52 wt-%, and 10.85 wt-%) on the same support material (Chromosorb W, HP, 80 to 100 mesh size) were investigated. In the molten state ($T>T_g$), the retention data of the column with the lowest loading (3.85 wt-%) significantly differed from the columns with higher polymer contents. This behavior may be understood in terms of a considerable surface adsorption in the case of the 3.85 wt-% loading. As the film thickness is increased, the rate of diffusion is no longer great enough to assure equilibrium during the passage of the probe through the column. Thus, to reach equilibrium conditions, the retention volume was measured at five carrier gas flow rates and extrapolated to zero. In order to minimize residual uncertainties, the same support material, at a polymer loading of approximately 8.5 wt-%, was used for all systems.

Consider a non-cristalline binary polymer system for which the glass transition temperatures of the pure components significantly differ. A homogenous blend shows only one glass transition, located between those of the pure polymers. Thus, at a given measurement temperature T, the distance to the blend glass transition T_g depends on the homopolymer concentration. If a series of blends differing in composition are to be compared, care must be taken to avoid artefacts resulting from the proximity of the various glass transition temperatures.

In Figure 1, ln V_g^0 of ethylbenzene in PPE is plotted against the reciprocal temperature. Well above and below the glass transition, the dependence is linear ([19]), where equilibrium surface adsorption (below T_g) and equilibrium sorption in the bulk and on the surface (above T_g) occur. The first deviation from linearity below T_g is observed at 220°C and is identified with the equilibrium glass transition temperature T_g. This agrees with the results of DeAraujo et al. ([200]) who found via DSC measurements T_g=220°C for a fractionated PPE sample with M_n=20,000 g/mol. Above the glass transition, the linear dependence of ln V_g^0 versus 1/T was observed at T>T_g+30°C. Thus, equilibrium sorption in PPE occurs at temperatures higher than 250°C. Since pure PS is not stable at these high temperatures, the retention data of the probes in PS measured between 130°C and 230°C were carefully extrapolated to T=290°C. The retention volumes of the corresponding probes in PPE were extrapolated to 220°C to get information about the sorption behavior of a pure PPE melt in this temperature region. With this technique a wide temperature range, which is not directly accessible, may be covered (Figure 2).

From 220° to 290°C, blends of different PS/PPE compositions were investigated. The free energy parameter χ_{23}' was calculated from Equation 6. The retention volumes of the probes in pure PS and PPE were obtained by the extrapolation procedure described above.

In all blend/solvent systems, a variation of χ_{23}' with the low molecular weight probe used was observed. In Table I, a representative example of the probe dependence is given. χ_{23}' varies in a 50 wt-% blend of PS and PPE at 220°C between -0.77 and -2.05. The most negative χ_{23}' was obtained from benzene, which is the best solvent for both polymers, whereas the non-solvent n-octane yielded a less negative χ_{23}'. Despite, this should not be a significant effect because all five selected probes are thermodynamically symmetric with regard to the pure polymers. Thus,

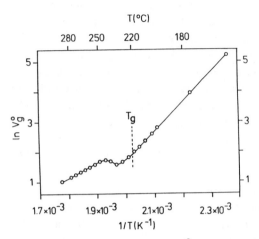

Figure 1. Retention diagram ln V_g^0 versus 1/T of ethyl
benzene in poly(2,6-dimethyl-1,4-phenylene oxide) (8.52
wt-% polymer loading).

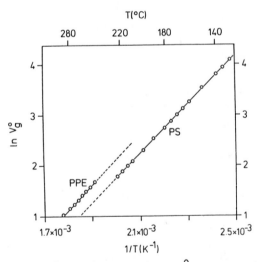

Figure 2. Retention diagram ln V_g^0 versus 1/T of ethyl
benzene in poly(2,6-dimethyl-1,4-phenylene oxide)(8.52
wt-%) and polystyrene (8.52 wt-%) in the molten state.

the known $\Delta\chi$-effect (21-23) was minimized and should not affect χ_{23}'. The results given here do not show any significant probe dependence. Certainly, this may also be due to the limited number of probes. For this reason, the χ_{23}' values from all probes were averaged.

Table I. Probe Dependence of χ_{23}' in
50 wt-% of PS/PPE at 220^0C

Probe	χ_{23}'
n-Octane	-0.77
Cyclohexane	-1.79
Benzene	-2.05
Toluene	-1.56
Ethyl benzene	-0.95
Average	-1.42

In Table II, some of the data are compared with very recent results of Fried and Su (24).

Table II. Comparision of χ_{23}' in a 50 wt-% Blend of
PS/PPE via IGC from Fried et al.[a](24) and
Present Paper[b]

Probe	Temperature (^{0}C)		
	260	270	280
Benzene[a]	-0.44	-0.36	-0.29
Benzene[b]	-1.48	-1.37	-1.23
Ethyl benzene[a]	-0.46	-0.39	-0.32
Ethyl benzene[b]	-0.58	-0.52	-0.46

χ_{23}' obtained from ethyl benzene agrees well with the present results, whereas χ_{23}' from benzene reasonably differs. Despite, there is no essential difference since in both studies, χ_{23}' is pronounced negative and increases with increasing temperature.

In Figures 3 and 4, χ_{23}' of four PS/PPE blends of different compositions are displayed as a function of the reciprocal temperature. χ_{23}' is negative, indicating thermodynamic miscibility. Nearly concentration independent straight lines with similiar slopes were obtained. Such a behavior was measured as well in mixtures of polystyrene and poly(styrene-co-p-bromo styrene) with small angle X-ray scattering (Koch, T. Diploma Thesis, University of Freiburg, Freiburg, 1987) and in blends of polycaprolactone and polyepichlorohydrin with IGC (25). Over a wide concentration range, an almost concentration independent χ parameter was obtained. Thus, a concentration dependence of χ in polymeric blends is not a natural necessity.

In Table III, results from this work and small angle neutron scattering data (SANS) of mixtures of d-PS in PPE and d-PPE in PS are given. Since χ_{23}' ($=V_1/V_2 \cdot \chi_{23}'$) is probe dependent, χ_{23}'/V_1 rather than χ_{23}' is used. Comparrison with χ/V_0 from SANS, where V_0 is the lattice site volume, shows that χ from IGC greatly differs from the SANS data.

Table III. Comparison of χ_{23}'/V_1 via IGC and
χ/V_0 from SANS in PS/PPE Blends

ϕ_{PS}	χ/V_0	T (^{0}C)
0.97	$-0.24*10^{-3}$	250[a]
0.03	$-0.30*10^{-3}$	glass[a]
0.90	$-0.36*10^{-3}$	glass[b]
0.05	$-0.18*10^{-3}$	glass[c]
0.55	$-0.18*10^{-3}$	200[d]
0.75[+]	$-5.50*10^{-3}$	250[e]
0.50[+]	$-8.20*10^{-3}$	250[e]
0.25[+]	$-8.00*10^{-3}$	250[e]

a)=Ref.(10); b)=Ref.(26); c)=Ref.(27); d)=Ref.(28);
e)= this paper; [+] $\chi/V_0 = \chi_{23}/V_1$.

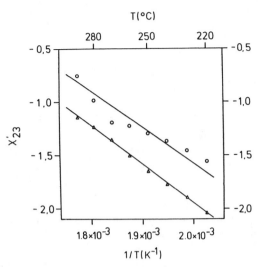

Figure 3. Temperature dependence of χ_{23}' in blends of polystyrene/poly(2,6-dimethyl-1,4-phenylene oxide) (o 75 wt-% PS/25 wt-% PPE, Δ 25 wt-% PS/75 wt-% PPE).

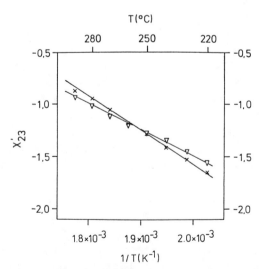

Figure 4. Temperature dependence of χ_{23}' in blends of polystyrene/poly(2,6-dimethyl-1,4-phenylene oxide) (* 15 wt-% PS/85 wt-% PPE, ∇ 50 wt-% PS/50 wt-% PPE.

This observation agrees with results of Walsh et al.
(29) who compared heat of mixing and IGC data. Thus, it
turns out that χ parameters from IGC may differ in more
than one order of magnitude from those obtained via scattering techniques and heat of mixing data. This finding
is confirmed in Table IV. IGC data are given for fifteen
compatible polymer pairs and χ_{23}' varies between -3.03
and +3.40. On the contrary, scattering techniques usually
yielded χ parameters between -0.02 and -0.1. Consequently, it seems to be a characteristic feature of IGC
in polymer blends to overestimate the Flory-Huggins free
energy parameter χ_{23}'.

With rising temperature, χ_{23}' linearly increased
(Figure 3 and 4). Linear extrapolation to χ_{23}'(critical),
where phase separation occurs,

$$\chi_{23}'(\text{critical}) = 0.5 \cdot [(V_1/V_2)^{0.5} + (V_1/V_3)^{0.5}]^2 \qquad (7)$$

lead to the phase diagram of the PS/PPE blend (Figure 5).
The error bars in Figure 5 indicate the uncertainty in
the phase diagram resulting from the extrapolation of
χ_{23}' to χ_{23}'(critical). As it can be seen from Equation 7,
χ_{23}'(critical) is temperature dependent. Nevertheless,
since this effect is small, the LCST is not significantly
affected. In the present case of high molecular weight
blends, χ_{23}'(critical) is approaching zero. The phase
diagram in Figure 5 reveals a LCST at approximately 360^0C
and agrees excellently with the prediction of Maconnachie
et al. (10). Another prediction of the LCST in PS/PPE can
be made from the data of Kramer et al. (28). They extracted χ = 0.145-78/T via forward recoil spectrometry
(FRES). For their system, χ(critical)= $+3.1*10^{-3}$, and
thus from the temperature dependence of χ a LCST of 276^0C
is predicted. This result is inconsistent with the data
of Ref.(10), (24), and the present paper.

At the end of this section, a comment has to be
given on the extrapolation procedure (see Figure 2),
which was applied to get information about a temperature
range that was not directly accessible. Within the mea-

Table IV. χ_{23}' of Fifteen Compatible
Polymer Blends via IGC

Component 1/ Component 2	Temperature range (^{0}C)	Concentration range (ϕ_1)	χ_{23}'
polystyrene/poly- (vinyl methyl ether)	30/50	0.35/0.65	-0.46/ -0.75[a]
	135/150	0.15/0.75	+0.16/ -1.64[b]
polystyrene/poly- (2,6-dimethyl-1,4- phenylene oxide)	240	0.25/0.85	+0.61/ -0.17[c]
	270	0.25/0.85	-1.36/ -1.12[d]
	260/280	0.50	-0.45/ -0.26[e]
poly(4-methyl styrene)/ poly(2,6-dimethyl-1,4- phenylene oxide)	270	0.50	-1.21[f]
poly(vinylidene fluo- ride)/poly(methyl methacrylate)	200	0.25/0.85	-0.58/ +0.19[g]
poly(vinyl chloride)/ di-n-octyl phtalate	110	0.82	-1.12[h]
poly(vinyl chloride)/ poly(ethyl methacrylate)	120	0.50	-3.03[i]
poly(vinyl chloride)/ poly(n-propyl meth- acrylate)	120	0.50	-1.52[i]

Continued on next page

Table IV. Continued

Component 1/ Component 2	Temperature range (^0C)	Concentration range (ϕ_1)	χ_{23}'
poly(vinyl chloride)/ poly(n-butyl meth acrylate)	120	0.50	-0.96^i
poly(vinyl chloride)/ poly(n-propyl acrylate)	120	0.50	$+0.38^i$
poly(vinylchloride)/ poly(n-butyl acrylate)	120	0.50	$+0.98^i$
poly(vinyl chloride)/ poly(ϵ-caprolactone)	120	0.10/0.90	$-2.60/$ $+0.05^j$
	120	0.50	$-0.40/$ $+1.16^k$
poly(vinyl chloride)/ poly(ethyl methacrylate)	80	0.25/0.75	$-1.62/$ -0.90^l
polyepichlorohydrin/ poly(methyl acrylate)	76	0.50	$-0.09/$ $+1.47^m$
	80/120	0.25/0.75	$-0.31/$ $+0.18^n$
polystyrene/poly(n-butyl methacrylate)	140	0.15/0.30	$-0.24/$ $+0.42^c$
chlorinated poly-(ethylene)/ethylene-vinyl acetate copolymer	70/100	0.25/0.75	$-1.20/$ $+3.40^o$

a=Ref.(30); b=Ref.(31); c=Ref.(32); d=present paper;
e=Ref.(24); f=Ref.(33); g=Ref.(34); h=Ref.(6);i=Ref.(35);
j=Ref.(36); k=Ref.(5); l=Ref.(37); m=Ref.(38);n=Ref.(25);
o=Ref.(29).

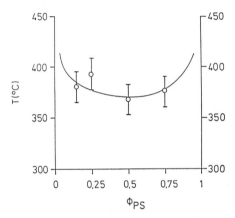

Figure 5. Phase diagram of polystyrene/poly(2,6-dimethyl-1,4-phenylene oxide) obtained via IGC.

surable range, a correlation coefficient usually better than 0.999 for ln V_g^0 versus 1/T was calculated. This high correlation justifies, at least in a first approximation, the applied extrapolation. In the case of pure PPE, additional uncertainties may be caused by chain fission and crosslinking at high temperatures.

In the other PS/PVME system, the complete temperature range of interest was accessible. The typical linear behavior of ln V_g^0 versus 1/T for PS with M_w=17,500 g/mol was obtained between 120^{0}C and 210^{0}C. Thus, all blends of PS/PVME and the pure PVME were measured in this temperature range. Five different probes (acetone, ethyl acetate, cyclohexane, n-octane, and ethyl benzene) were used to get χ_{23}'. As observed in the PS/PPE blends, χ_{23}' scattered depending on the probe. Because of the limited number of probes and their different chemical nature, no systematical probe dependence could be detected. Thus, the different χ_{23}' values were averaged.

In Figure 6, χ_{23}' of four PS/PVME blends is displayed as a function of the reciprocal temperature. For the blends, a linear relation of χ_{23}' versus 1/T was observed. Contrary to the PS/PPE system, χ_{23}' varies significantly with the homopolymer concentration in the blend. The 50 wt-% mixture showed small positive χ_{23}', due to the proximity to the LCST. Blends of high contents of either one component were far away from their corresponding demixing temperatures, and thus, showed pronounced negative χ_{23}'. Since the slopes of χ_{23}' versus 1/T vary, the enthalpic contribution to χ_{23}' must also be a function of the concentration.

Extrapolation of χ_{23}' to χ_{23}'(critical) results in the phase diagram of the PS/PVME system (Figure 7). From the concentration dependence of χ_{23}' (Figure 6), the critical point is ϕ(critical)≈0.5 and T(critical)≈130^{0}C, whereas the demixing temperatures of the other blend ratios are shifted toward higher temperatures. The phase diagram of PS/PVME, obtained via IGC, agrees with turbidity measurements from literature (9).

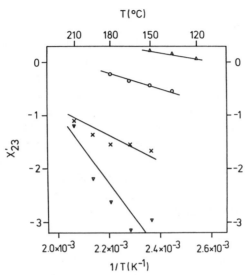

Figure 6. Temperature dependence of χ_{23}' in blends of polystyrene/poly(vinyl methyl ether) (x 15 wt-% PS/85 wt-% PVME, ▼ 25 wt-% PS/75 wt-% PVME, ▲ 50 wt-% PS/50 wt-% PVME, o 75 wt-% PS/25 wt-% PVME).

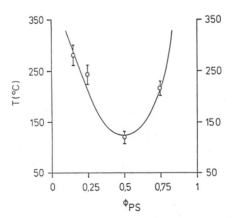

Figure 7. Phase diagram of polystyrene/poly(vinyl methyl ether) obtained via IGC.

Conclusions

Two polymer blends, PS/PVME and PS/PPE, were investigated. For all blend compositions, negative χ_{23}' parameters were found, indicating thermodynamic compatibility. Through comparison with small angle neutron scattering results and IGC literature data, it turned out that IGC often overestimates the χ_{23}' parameter. Despite this discrepancy, a correct qualitative interpretation of polymer blends via IGC can be given. χ_{23}' depends linearly on the reciprocal temperature. From this relation, the phase diagrams of both blend systems were obtained. IGC enables a consistent and qualitative thermodynamic characterization, and even a quantitative description of phase diagrams in polymeric blends.

Acknowledgments: We thank the Deutsche Forschungsgemeinschaft, Sonderforschungsbereich SFB 60, for financial support of the present work. S.K. would like to thank Clara C. Pizaña for editorial suggestions.

Literature Cited

1.   Olabisi, O.; Robeson, L. M.; Shaw, M. T. Polymer-
      Polymer Miscibility; Academic Press: New York,
      1979.
2.   Paul, D. R.; Newman, S. Polymer Blends; Academic
      Press: New York, 1978; Vol. I, II.
3.   Scott, R. L. J. Chem. Phys. 1949, 17, 279.
4.   Deshpande, D. D.; Patterson, D.; Schreiber, H. P.;
      Su, C. S. Macromolecules 1974, 7, 530.
5.   Olabisi, O. Macromolecules 1975, 8, 316.
6.   Su, C. S.; Patterson, D.; Schreiber, H. P. J. Appl.
      Polym. Sci. 1976, 20, 1025.
7.   Stoelting, J.; Karasz, F. E.; MacKnight, W. J.
      Polym. Eng. Sci. 1970, 10, 133.
8.   Bank, M.; Leffingwell, J.; Thies, C. Macromolecules
      1971, 4, 43.
9.   Nishi, T.; Kwei, T. K. Polymer 1975, 16, 285.

10. Maconnachie, A.; Kambour, R. P.; White, O. M.; Rostami, S.; Walsh, D. J. Macromolecules 1984, 17, 2645.

11. Leung, Y. K.; Eichinger, B. E. J. Phys. Chem. 1974, 78, 60.

12. Deshpande, D. D.; Tyagi, O. S. Macromolecules 1978, 11, 746.

13. Littlewood, A. B.; Phillips, G. S. G.; Price, D. T. J. Chem. Soc. 1955, 1480.

14. Patterson, D.; Tewari, Y. B.; Schreiber, H. P. J. Chem. Soc., Faraday Trans. 2 1972, 68, 885.

15. Dreisbach, D. R. Adv. Chem. Ser. 1959, Vol. 15, 22, 29.

16. Timmermans, J. Physico-Chemical Constants of Pure Organic Compounds; Elsevier: New York, 1950; Vol. I; Ibid.; 1965; Vol. II.

17. Höcker, H.; Flory, P. J. Trans. Faraday. Soc. 1971, 67, 2270.

18. Zoller, P.; Hoehn, H. H. J. Polym. Sci. Phys. Ed. 1982, 20, 1385.

19. Braun, J. M.; Guillet, J. E. Macromolecules 1975, 8, 882.

20. DeAraujo, M. A.; Stadler, R. Makromol. Chem. 1988, in press.

21. Zeman, L.; Patterson, D. Macromolecules 1972, 5, 513.

22. Hsu, C. C.; Prausnitz, J. M. Macromolecules 1974, 7, 320.

23. Robard, A.; Patterson, D.; Delmas, G. Macromolecules 1977, 10, 706.

24. Fried, J. R.; Su, A. C. Proc. ACS Div. Polym. Mat.: Sci.& Eng. 1988, 58, 928.

25. El-Hibri, M. J.; Cheng, W.; Munk, P. Proc. ACS Div. Polym. Mat.: Sci.& Eng. 1988, 58, 741.

26. Kambour, R. P.; Bopp, R. C.; Maconnachie, A.; MacKnight, W. J. Polymer 1980, 21, 133.

27. Maconnachie, A.; Kambour, R. P.; Bopp, R. C. Polymer 1984, 25, 357.

28. Composto, R. J.; Mayer, J. W.; Kramer, E. J.; White, D. M. Phys. Rev. Lett. 1986, 57, 1312.

29. Walsh, D. J.; Higgins, J. S.; Rostami, S.;
     Weeraperuma, K. Macromolecules 1983, 16, 391.
30. Kwei, T. K.; Nishi, T.; Roberts, R. F.
     Macromolecules 1974, 7, 667.
31. Klotz, S.; Schuster, R.; Cantow, H.-J. Makromol.
     Chem. 1986, 187, 1491.
32. DiPaola-Baranyi, G. Proc. ACS Div. Polym. Mat.:
     Sci.& Eng. 1988, 58, 735.
33. Su, A. C.; Fried, J. R. In Multicomponent Polymer
     Materials; Paul, D. R.; Sperling, L. H., Eds.;
     Advances in Chemistry Series No. 211; American
     Chemical Society: Washington, DC, 1986; p 59
34. DiPaola-Baranyi, G.; Fletcher, S. J.; Degre, P.
     Macromolecules 1982, 15, 885.
35. Walsh, D. J.; McKeown, J. G. Polymer 1980, 21,
     1335.
36   Riedl, B.; Prud'homme, R. R. ACS Polym. Prepr. Div.
     Polym. Chem., Inc. 1987, 28(2), 138.
37. Zhikuan, C.; Walsh, D. J. Eur. Polym. J. 1983,
     19, 519.
38. Al-Saigh, Z.; Munk, P. Macromolecules 1984, 17,
     803.

RECEIVED November 2, 1988

Chapter 12

Interaction Parameters
of Poly(2,6-dimethyl-1,4-phenylene oxide) Blends

A. C. Su[1] and J. R. Fried

Department of Chemical and Nuclear Engineering and the Polymer
Research Center, University of Cincinnati, Cincinnati, OH 45221

Specific retention volumes of five probes in
polystyrene (PS), poly(4-methylstyrene) (P4MS),
poly(2,6-dimethyl-1,4-phenylene oxide) (PMMPO),
and 50/50 blends of PMMPO/PS and PMMPO/P4MS
were determined at temperatures between 200 and
280°C. Loading determinations were made according to the soaking method of Al-Saigh and Munk.
In each case, a correction to the data was made
for a small contribution to probe retention by
the uncoated dimethyldichloro-silane (DMCS)-
treated packing. Values of the apparent Flory
interaction parameter, χ, calculated for the
PMMPO/PS and PMMPO/P4MS blends were both negative. In agreement with conclusions from earlier thermal analysis and mechanical property
studies, comparison of relative values indicate
that the miscibility of the PMMPO/P4MS pair is
more marginal, that is, χ is less negative,
than that of PMMPO/PS.

As comprehensively reviewed by Lipson and Guillet (1),
inverse gas chromatography (IGC) has been used as a
convenient tool to study the thermodynamic properties of
polymeric systems. Despite its wide usage, all experimental and theoretical factors in this technique are not
fully understood. Loading determination, usually done by
means of extraction or calcination, has been considered
to be the most significant source of experimental error
(2). Other factors, such as concentration effects associated with large injection sizes, slow diffusion of solute
probe molecules in the stationary phase, and adsorption
of probes onto the liquid-support interface, may also af-

---

[1]Current address: Institute of Materials Science, National Sun Yat-Sen University, Kaohsiung,
Taiwan 80424, Republic of China

0097–6156/89/0391–0155$06.00/0
© 1989 American Chemical Society

fect the measured retention volume. Typical procedures
aimed at minimizing errors from the latter sources include, respectively, the use of an arbitrarily chosen
small injection size (or extrapolation of retention volumes measured using different injection sizes to zero injection size), the use of low loadings or slow carrier
gas flow rates, and the extrapolation of specific retention volumes obtained in columns of different loadings to
infinite polymer loadings.

In an effort to eliminate errors from loading
determination, Al-Saigh and Munk (3) used a "soaking"
procedure in which the stationary phase was loaded onto
the support by repeatedly dropping a small amount of the
polymer solution onto the piled support particles, allowing the solvent to evaporate, and then remixing the particles. In this way, the polymer loading could be accurately established. With the error from loading determination minimized, Munk et al. (4) were able to show that,
even after dimethyldichlorosilane (DMCS) treatment, the
residual adsorption sites on the support surface may
still contribute significantly to the probe retention,
especially when the injection size is small. Furthermore,
they observed a linear contribution to the probe retention, which they attributed to a small amount of unexpected polymer, presumably polydimethylsiloxane (PDMS)
deposited on the support surface during the DMCS treatment. This linear contribution became more significant
when the column loading was low. These observations suggest that some of the commonly used procedures adopted
for reducing experimental errors may actually introduce
further inaccuracies. Assuming that the surface adsorption and parallel retention contributions to probe retention are the same for loaded and unloaded columns, Munk
et al. subtracted retention volumes for the uncoated supports from the retention volumes determined for the
loaded columns. By this method, retention volumes should
be independent of probe injection size, carrier gas flow
rate, and polymer loading.

A previous communication (5) reported a preliminary
IGC determination of the Flory interaction parameter, χ,
for blends of poly(2,6-dimethyl-1,4-phenylene oxide)
(PMMPO) with polystyrene (PS) and with poly(4-methyl-
styrene) (P4MS) at 270°C. Results of earlier differential
scanning calorimetry, density, and mechanical property
measurements (6) suggested that miscibility of PMMPO with
P4MS was slightly more marginal than with PS. This was
not confirmed by our preliminary IGC results, which
yielded a strongly negative χ value for PMMPO/P4MS
compared to a slightly positive value for PMMPO/PS. The
reason for this discrepancy was not clear at the time. In
view of the recent work of Munk et al. (4), experimental
errors in the IGC procedures may have been responsible
for this discrepancy; therefore, the earlier work was
repeated and expanded following, in part, their
experimental procedures.

Materials and Methods

Materials used in this study and details of data analysis
have been described in an earlier communication (5). Only
essential details and differences between the two studies
are given below.
      As in the earlier study, the support was Chromosorb
P. This packing is a commercially available (Johns-
Manville), acid-washed, and dimethyldichlorosilane
(DMCS)-treated calcinated diatomite with a nominal size
of 60 to 80 mesh. The DMCS treatment is commonly used to
minimize specific adsorption of the solute probes on the
uncoated packing. Instead of the traditional slurry procedure used previously, packings were coated by the soaking method of Al-Saigh and Munk (3) with a slight modification. In place of a watch glass, as used by Al-Saigh
and Munk, a piece of folded paper was used to hold the
packing during soaking. A small quantity of the polymer
solution, 5 wt/vol-% in chloroform, was dropped by
pipette onto the support material. Typically, the solution penetrated 1 to 2 mm into the center of the packing
material. This was left to dry and then the packing was
mixed by manually deforming the paper. This wetting procedure was repeated until all the solution was used. This
procedure took approximately five hours. A blank column,
containing no polymer support, was similarly prepared for
the purpose of correcting contributions of probe retention from the support. Characteristics of the columns
used in the present study are given in Table I.

Table I. Column Loading Data

Column	Loading Wt, g	Support Wt, g	Loading %	Extracted Polymer, g	Recovery %
Blank 1	0	10.008	0	–	–
Blank 2	0	10.310	0	0.002	–
PS	0.718	9.858	7.29	0.705	98.2
P4MS	0.635	8.982	7.07	0.116	18.3
PMMPO 1	0.705	10.044	7.02	0.036	5.1
PMMPO 2	0.680	9.936	6.84	0.323	47.5
PMMPO 3	0.749	10.065	7.44	0.755	100.8
PMMPO/PS[a]	0.658	9.234	7.12	0.272	41.4
PMMPO/P4MS[b]	0.680	9.774	6.96	0.034	4.9

a. 50.1/49.9 wt/wt
b. 50.2/49.8 wt/wt

      The carrier gas was helium; flow rate was approxi-
mately 5 mL/min, as measured by a homemade 9 mL bubble
flow meter. Because of its small size, the helium diffusional loss, as discussed by Munk et al. (7), was deter-

mined to be negligible at this flow rate (8). Solvent
probes were reagent-grade benzene, toluene, ethylbenzene,
chlorobenzene, and bromobenzene. Typical injection sizes
were 6-15 μL for air and less than 0.06 μL for solutes.
At least four injections of different amounts of the
probe were made at each temperature. Typically, steady-
state temperature was reached within two hours and measurements were completed within four hours.

To minimize the contribution of adsorption of the
solute probe onto the support surface of the packing,
retention times were first extrapolated to infinite injection size by plotting retention time against peak
height. Since injection sizes were small, this procedure
is unlikely to introduce significant deviations from the
ideal limit of infinite dilution. In the present study,
the contribution of adsorption varied from column to column. This was considered to result from variations in
packing breakage, and therefore surface area, during the
soaking procedure.

IGC data are plotted as log specific retention volume versus reciprocal temperature, where the specific retention volume (V_g) is calculated as (1)

$$V_g = t_N FJ/W \tag{1a}$$

In Equation 1a, W is the weight of the stationary phase
and t_N is the net retention time defined as

$$t_N = t_R - t_M \tag{1b}$$

where t_R is the recorded time for the solute peak and t_M
is the time for the marker (air) peak. The carrier flow
rate, F, is corrected for water vapor pressure (P_w) and
standardized to $0°C$ according to

$$F = (273.16/T_a)Q(P_o - P_w)/P_o \tag{1c}$$

where Q is the volumetric flow rate of the carrier gas
(helium) measured at the column outlet by a soap bubble
flow meter at ambient temperature (T_a) and 1 atm pressure. Pressure drop across the column is incorporated in
the correction factor, J, as

$$J = 3[(P_i/P_o)^2 - 1]/2[(P_i/P_o)^3 - 1] \tag{1d}$$

where P_i and P_o are the pressures at the inlet and outlet
of the column.

Apparent χ values for each blend, $\chi_{23,app}'$, are
given as (3)

$$\chi_{23,app}' = \chi_{23}(V_1/V_2) = \{\ln[V_{g,b}/(w_2\nu_2 + w_3\nu_3)] -$$

$$\phi_2\ln(V_{g,2}/\nu_2) - \phi_3\ln(V_{g,3}/\nu_3)\}/\phi_2\phi_3 \tag{2}$$

where V is molar volume, w is weight fraction, ν is specific volume at the measurement temperature, and ϕ is volume fraction; subscripts 1, 2, 3, and b refer to the solute probe, polymer 2, polymer 3, and the blend, respectively.

Results and Discussion

Retention diagrams of the solute probes on the unloaded columns are shown in Figure 1. Data were fitted by linear regression. Comparison of data for benzene, toluene, and ethylbenzene at 120, 140, 160, and 180°C with published specific volumes of these probes in a PDMS stationary phase (9) indicates a value of 0.24 ± 0.03 wt-% hydrocarbon (methyl groups) on the support. This compares favorably with a value of 0.2 wt-% hydrocarbons obtained by elemental analysis by Al-Saigh and Munk (4) for their packing. By using chloroform in the Soxhlet extraction of the packing, 1.5 mg/g packing or 0.15 wt-% was recovered (blank column 2, Table I).

Specific retention volumes and calculated χ_{12} values for the three sets of polymer-probe pairs are given in Tables II to IV. In all cases, specific retention volumes were corrected for support retention. Retention diagrams for polystyrene and PMMPO are shown in Figures 2 to 3.

In the case of PMMPO, it was observed that V_g increased with time of heating. For example, values that were initially 2.23 and 1.95 mL/g at 270 and 280°C increased to 4.93 and 3.52, respectively, when the same column was used for additional measurements over one month's time. The cause of this observed increase is not certain; however, PMMPO is believed to undergo branching or crosslinking in the melt at elevated temperatures (10). Poly(4-methylstyrene) has also been reported to undergo crosslinking at temperatures above 250°C by means of transfer reactions to the p-methyl group (11). To minimize exposure time to high temperatures, two different PMMPO columns (#1 and #2) were used for IGC measurements and data at only two temperatures were used for each. Agreement between values of V_g obtained at 270°C for each of the two columns was good as shown in Figure 3.

Comparison of the specific retention volumes for the probes - benzene, toluene, and ethylbenzene - that are common between the present and earlier study indicated significant differences in results for the P4MS and PMMPO columns. Only values for the PS column were comparable. In the previous study, Soxhlet extraction was used to determine column loading. Results of Soxhlet chloroform extraction of the columns used in the present study are included in Table I. These results indicate that nearly total recovery of the coating polymer was obtained only for PS and for PMMPO, column 3, which was coated but not heated in the chromatograph. The recovery of polymer from PMMPO columns 1 and 2, P4MS, and the blend columns was incomplete. Since P4MS and PMMPO both have aromatic

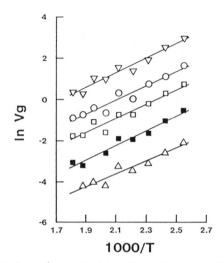

1000/T

Figure 1. Retention plots of probes on the unloaded support (blank 1, Table I). Specific retention volumes, V_g, are in units of mL/g-support. Data points for benzene (△) as shown. Data for toluene (■), ethylbenzene (□), chlorobenzene (o), and bromobenzene (▽) are shifted vertically along the ordinate by units of 1, 2, 3, and 4 (log V_g), respectively, to facilitate comparison.

Table II. Specific Retention Volumes,[a] χ Values, and
Linear Regression Parameters[b] for the Polystyrene Column

T, °C	Benzene V_g	χ_{12}	Toluene V_g	χ_{12}	Ethyl-benzene V_g	χ_{12}	Chloro-benzene V_g	χ_{12}	Bromo-benzene V_g	χ_{12}
200	4.28	0.379	6.55	0.403	9.59	0.413	11.09	0.358	18.57	0.329
210	3.65	0.384	5.58	0.393	7.79	0.438	9.21	0.363	15.01	0.347
220	3.13	0.387	4.66	0.410	6.33	0.469	7.50	0.397	12.33	0.357
230	2.84	0.343	4.07	0.390	5.50	0.441	6.29	0.409	10.08	0.382
240	2.38	0.381	3.38	0.426	4.64	0.447	5.42	0.399	8.56	0.377
250	2.04	0.397	2.87	0.446	3.85	0.479	4.50	0.433	6.99	0.419
260	2.14	0.258	2.88	0.302	3.74	0.357	4.34	0.321	6.57	0.326
270	1.82	0.227	2.38	0.356	3.13	0.392	3.68	0.346	5.64	0.329
A	3.111		3.689		4.042		4.045		4.405	
B	-5.151		-5.933		-6.319		-6.170		-6.417	

a. V_g, mL/g-coating
b. ln V_g = A(1000/T) + B, where T is in K

Table III. Specific Retention Volumes[a], χ Values, and Linear Regression Parameters[b] for the Poly(4-methylstyrene) Column

T,°C	Benzene V_g	χ_{12}	Toluene V_g	χ_{12}	Ethyl-benzene V_g	χ_{12}	Chloro-benzene V_g	χ_{12}	Bromo-benzene V_g	χ_{12}
200	4.24	0.423	6.65	0.423	9.60	0.448	11.23	0.381	18.57	0.365
210	3.70	0.405	5.64	0.418	7.96	0.451	9.37	0.382	15.08	0.378
220	3.17	0.411	4.78	0.421	6.63	0.460	7.95	0.375	12.63	0.370
230	2.91	0.354	4.25	0.382	5.83	0.419	6.72	0.379	10.50	0.378
240	2.70	0.292	3.76	0.355	4.86	0.439	6.00	0.335	9.13	0.348
250	2.14	0.385	3.17	0.383	4.30	0.406	5.10	0.343	7.69	0.361
260	1.67	0.494	2.30	0.567	3.01	0.611	3.68	0.524	5.45	0.551
270	1.61	0.386	2.26	0.448	3.00	0.474	3.55	0.419	5.52	0.388
A	3.664		4.094		4.431		4.335		4.654	
B	-6.260		-6.729		-7.085		-6.719		-6.906	

a. V_g, mL/g-coating
b. $\ln V_g = A(1000/T) + B$, where T is in K

Table IV. Specific Retention Volumes[a], χ Values, and Linear Regression Parameters[b] for the Poly(2,6-dimethyl-1,4-phenylene oxide) Columns

T,°C	Benzene V_g	χ_{12}	Toluene V_g	χ_{12}	Ethyl-benzene V_g	χ_{12}	Chloro-benzene V_g	χ_{12}	Bromo-benzene V_g	χ_{12}
270[c]	2.23	-0.030	2.96	0.088	3.94	0.111	4.76	0.036	7.15	0.041
270[d]	2.29	-0.058	2.97	0.084	3.76	0.158	4.53	0.086	6.55	0.127
275[c]	2.13	-0.061	2.73	0.101	3.46	0.170	4.17	0.100	6.17	0.116
280[d]	1.95	-0.061	2.45	0.143	3.16	0.194	3.70	0.154	5.48	0.163
A	4.428		5.636		5.964		6.751		6.628	
B	-7.333		-9.288		-9.634		-10.893		-10.277	

a. V_g, mL/g-coating
b. $\ln V_g = A(1000/T) + B$, where T is in K
c. Column 1 (see Table I)
d. Column 2 (see Table I)

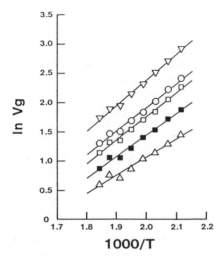

Figure 2. Retention plots of probes on polystyrene.
Data points are: (△) benzene; (■) toluene; (□) ethylbenzene; (o) chlorobenzene; and (▽) bromobenzene. Data
points represent actual, unshifted values. Lines represent linear regression fit of data.

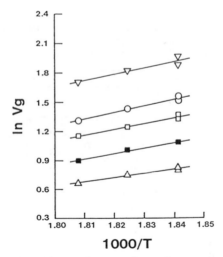

Figure 3. Retention plots of probes on poly(2,6-
dimethyl-1,4-phenylene oxide), columns 1 and 2. Legend
same as in Figure 2.

methyl groups, it is probable that the non-extractable portion of the loading may be due to the formation of methylene bridges between P4MS or PMMPO with partially reacted DMCS on the packing surface at high temperature. The reactive group is probably Si-Cl, which results when DMDC reacts with a lone silanol group on the packing surface (12). Similar results for tetramethyl bisphenol-A polycarbonate were observed (8).

Specific retention volumes and regression parameters for the PMMPO/PS and PMMPO/P4MS blends are given in Tables V and VI, respectively. Apparent x_{23} values calculated for PMMPO/PS and for PMMPO/P4MS blends at 260, 270, and 280°C are given in Table VII. Values used for V_g in Equation 2 were obtained by linear regression (and extrapolation when necessary) of the V_g data (Tables II to VI). Apparent x_{23} values were calculated only for 260, 270, and 280°C because of the limited temperature range of the PMMPO retention data.

The negative $x_{23,app}'$ values are in agreement with the observed compatibility of these blends. The temperature dependence of $x_{23,app}'$ for the PMMPO/PS blend suggests the occurrence of a lower critical solution temperature (LCST) above 300°C in agreement with the conclusion drawn from small-angle neutron scattering measurements (13). In contrast, the temperature dependence of $x_{23,app}'$ for the PMMPO/P4MS blend is rather weak, exhibiting stronger probe-to-probe variations. Comparison of $x_{23,app}'$ between the two blends indicates a slightly more favorable interaction in the PMMPO/PS pair in agreement with earlier glassy state property measurements (6).

Table V. Specific Retention Volumes[a] and Linear Regression Parameters[b] for the (50.1/49.9) PMMPO/PS Blend Column

T, °C	Benzene V_g	Toluene V_g	Ethyl-benzene V_g	Chloro-benzene V_g	Bromo-benzene V_g
200	4.57	6.96	9.85	11.68	19.23
210	3.77	5.64	8.00	9.58	15.63
220	3.27	4.76	6.42	7.91	12.72
230	2.77	4.02	5.36	6.70	10.83
240	2.49	3.51	4.74	5.70	8.94
250	2.15	3.19	4.12	4.88	7.71
260	2.06	2.65	3.51	4.23	6.36
270	1.86	2.42	2.94	3.72	5.59
A	3.295	3.843	4.303	4.204	4.546
B	-5.488	-6.216	-6.835	-6.444	-6.661

a. V_g, mL/g-coating
b. $\ln V_g = A(1000/T) + B$, where T is in K

Table VI. Specific Retention Volumes[a] and Linear Regression
Parameters[b] for the (50.2/49.8) PMMPO/P4MS Blend Columns

T, ^{o}C	Benzene V_g	Toluene V_g	Ethyl- benzene V_g	Chloro- benzene V_g	Bromo- benzene V_g
210	4.75	7.94	11.99	13.40	21.85
220	4.01	6.19	9.36	10.41	16.48
230	3.26	4.79	7.09	8.09	13.03
240	2.85	4.15	5.88	6.69	10.58
25C	2.48	3.49	4.73	5.36	8.38
260	2.18	2.98	3.92	4.57	6.90
270	1.83	2.55	3.35	3.89	5.24
A	4.087	4.912	5.614	5.421	6.135
B	-6.910	-8.129	-9.160	-8.653	-9.610

a. V_g, mL/g-coating
b. ln V_g = A(1000/T) + B, where T is in K

Table VII. Apparent Interaction Parameters, $\chi_{23,app}'$, for
PMMPO/PS and PMMPO/P4MS Blends [a]

PMMPO/PS Probe	260°C	270°C	280°C
benzene	-0.44	-0.36	-0.29
toluene	-0.42	-0.31	-0.25
ethylbenzene	-0.46	-0.39	-0.32
chlorobenzene	-0.50	-0.36	-0.23
bromobenzene	-0.44	-0.33	-0.22
average χ_{23}[b]	-0.45 ± 0.03	-0.35 ± 0.03	-0.26 ± 0.04

PMMPO/P4MS Probe	260°C	270°C	280°C
benzene	-0.09	-0.10	-0.09
toluene	-0.08	-0.09	-0.09
ethylbenzene	-0.04	-0.08	-0.12
chlorobenzene	-0.23	-0.26	-0.21
bromobenzene	-0.28	-0.33	-0.38
average χ_{23}[b]	-0.15 ± 0.11	-0.17 ± 0.12	-0.18 ± 0.12

a. Normalized to a repeat unit of PMMPO.
b. Error bounds correspond to standard deviations of the mean
value for different probes.

Literature Cited

1. Lipson, J. E. G.; Guillet, J. E. In Developments in Polymer Characterization; Dawkins, J. V., Ed.; Applied Science: London, 1982; Vol. 3, Chapter 2.
2. Laub, R. J.; Purnell, J. H.; Williams, P. S.; Harbison, W. P.; Martire, D. E. J. Chromatogr. 1978, 155, 233.
3. Al-Saigh. Z. Y.; Munk, P. Macromolecules 1984, 17, 803.
4. Card, T. W.; Al-Saigh, Z. Y.; Munk, P. Macromolecules 1985, 18, 1030.
5. Su, A. C.; Fried, J. R. In Multicomponent Polymer Materials; Paul, D. R.; Sperling, L. H., Eds.; Advances in Chemistry Series No. 211; American Chemical Society: Washington, DC, 1986; p 59.
6. Fried, J. R.; Lorenz, T.; Ramdas, A. Polym. Eng. Sci. 1985, 25, 1048.
7. Card, T. W.; Al-Saigh, Z. Y.; Munk, P. J. Chromatogr. 1984, 301, 261.
8. Su, A. C. Ph.D. Thesis, University of Cincinnati, Cincinnati, 1986.
9. Galin, M. Macromolecules 1977, 10, 1239.
10. Singh, R. P. Eur. Polym. J. 1982, 18, 117.
11. Schroder, U. K. O. Makromol. Chem. 1987, 188, 2775.
12. Boheman, J.; Langer, S. H.; Perrett, R. H.; Purnell, J. H. J. Chem. Soc. (London) 1960, 2444.
13. Maconnachie, A.; Kambour, R. P.; White, D. M.; Rostami, S.; Walsh, D. J. Macromolecules 1984, 17, 2645.

RECEIVED November 2, 1988

SURFACE AND INTERFACE
CHARACTERIZATION

Chapter 13

Properties of Carbon Fiber Surfaces

Aleksandar J. Vukov and Derek G. Gray

Pulp and Paper Research Institute of Canada and Department of
Chemistry, McGill University, 3420 University Street, Montréal, Québec
H3A 2A7, Canada

Surface property studies of high-strength and high-
modulus carbon fibers indicate that they possess
higher surface free energies than previously
believed. London dispersion components of the surface free energy were calculated from the increment
per methylene group in the free energy of adsorption
of n-alkanes at zero coverage. Values typical of
low-energy surfaces were obtained for both types of
"as received" carbon fibers. Cleaning of the fibers
by pretreating at elevated temperatures under nitrogen caused a significant increase in the London component. This was attributed to the desorption of
physically adsorbed species (CO_2, H_2O) that occupied
the high energy sites on "as received" fibers.
Similar results were obtained in the finite coverage
region where London components were calculated from
the spreading pressures of the hydrocarbons.
Brunauer-Deming-Deming-Teller type II adsorption
isotherms were measured for n-alkanes on carbon
fibers. The fibers were pretreated by heating to
various temperatures under nitrogen. The BET surface
areas of the fibers increased with increasing pretreatment temperatures, because of the presence of
microporosity.

Carbon fibers constitute the major load-bearing element of carbon
fiber reinforced plastics. The ability of these composites to use
effectively the strength and stiffness of carbon fibers depends
upon the strength of the interfacial zone, which is closely
related to the surface free energies of the carbon fibers and
matrix.
The values of the surface free energy for the polymer matrix
can be obtained from classical contact angle measurements. In the
case of fibers, surface roughness and the presence of surface

0097–6156/89/0391–0168$06.00/0
© 1989 American Chemical Society

energy gradients often result in contact angle hysteresis. Some results obtained by contact angle measurements are reported in the literature (1-3). Schultz et al. (4) have recently used a new method with two immiscible liquid phases that makes it possible to measure finite contact angles. While the carbon fibers with a polymeric coating gave reproducible results typical for the polymer, fibers without coating gave poor results. The observed hysteresis effect is probably due to surface energy gradients.

In this paper, an alternative route to estimate the surface energy of carbon fibers is presented, based not on contact angle measurements, but on inverse gas chromatography (IGC). The method is, in principle, applicable to any polymer, but the practical difficulties inherent in differentiating between surface adsorption and bulk sorption of vapours have restricted applications to substrates in which the probe vapour is essentially insoluble. Two distinct sets of IGC experiments may be performed on surfaces: approaching "zero coverage", where the adsorption isotherm is essentially linear, and at "finite concentration", where the shape of the adsorption isotherm reflects the build-up of multilayers on the surface. Results in these two regions are complementary and may be combined with wetting experiments to develop a detailed picture of the surface properties.

Zero Coverage. Inverse gas chromatography has been used successfully in the past decade for studying the surface properties of solids by adsorption of vapour at a gas-solid interface. Unlike conventional adsorption techniques, IGC allows the measurement of adsorption data down to low vapour concentrations where the surface coverage approaches zero, adsorbate-adsorbate interactions are negligible, and thermodynamic functions depend on only adsorbate-adsorbent interactions.

IGC has been used at zero surface coverage to characterize the surfaces of cellulose (5), cellophane (6), and poly(ethylene terephthalate) film (7). Surface properties of intact textile fibers were also studied by IGC (8). Domingo-Garcia et al. (9) have recently characterized graphite and graphitized carbon black surfaces with this method, and some zero coverage results on carbon fibers have appeared (10).

The fundamental parameter measured in gas chromatography is the retention volume, which is the volume of carrier gas required to elute a zone of solute vapour. For surface adsorption,

$$V_N = K_s A \qquad (1)$$

where V_N is the measured net retention volume of the probe corrected for pressure drop and column temperature, A is the total surface area of the stationary phase, and K_s is the surface partition coefficient. From K_s and its temperature dependence, thermodynamic data describing the retention process may be derived. For example, the standard free energy change, ΔG_A^o, for the isothermal adsorption of 1 mol of adsorbate, from the standard gaseous state to a standard state on the surface, is given by Equation 2:

$$\Delta G_A^O = -RT \ln (K_s \cdot p_{s,g}/\pi_s) \qquad (2)$$

where $p_{s,g}$ is the adsorbate vapour pressure in the gaseous standard state, π_s is the vapour pressure in equilibrium with the standard adsorption state, R is the gas constant, and T is the column temperature. The standard reference states are taken as $p_{s,g} = 101 \; kN \cdot m^{-2}$ (1 atm) and $\pi_s = 0.338 \; mN \cdot m^{-1}$. The latter value, proposed by de Boer (11), arbitrarily defines the standard surface pressure as the pressure at which the average distance of separation between molecules in the adsorbed state equals that in the standard gas state.

The differential heat of adsorption of the probe, q_d, may be obtained from the temperature dependence of K_s, according to

$$d(\ln K_s) / d(1/T) = q_d/R \qquad (3)$$

Provided that q_d is temperature independent, Equation 3 predicts a linear relationship between $\ln K_s$ and T^{-1}.

Finite Concentration. In this concentration range, surface adsorption results in nonlinear isotherms in which partition coefficients and retention volumes are dependent upon the adsorbate concentration in the gas phase. This means that a single partition coefficient, K_s (= Γ/c), is insufficient to characterize the process and the differential $(\partial\Gamma/\partial c)_T$ is required. Here, Γ is the surface excess of adsorbate expressed in $mol \cdot m^{-2}$, c is the gas phase adsorbate concentration, and T is the column temperature. Nonlinear isotherms give asymmetric peaks, whose shapes and retention volumes depend on the concentration of the probe.

The dependence of the retention volume on the adsorbate concentration in the gas phase has proved to be a useful and rapid way to determine adsorption isotherms (12). The adsorption of organic molecules and water on glassy polymers (13), cellulose fibers, paper (14-16), cellophane (17), glass fibers (18), textile fibers (8), and carbons (19) has been measured by IGC. The net retention volume V_N, corrected for the pressure drop across the column by the gas compressibility factor of James and Martin (20) is given by

$$V_N = (d\Gamma/dc) \cdot A = (dq/dc) \cdot w = RT \; (dq/dp) \cdot w \qquad (4)$$

where A is the total surface area of adsorbent, q is the surface excess of probe expressed in $mol \cdot g^{-1}$, w is the total weight of adsorbent, p is the vapour pressure of probe, and R is the gas constant. The adsorption isotherm is obtained by integrating Equation 4 so that

$$q = (1/w \; R \; T \;) \cdot \int V_N \; dp \qquad (5)$$

Thermodynamic data may be determined as a function of surface coverage from the temperature variation of the adsorption isotherms. The isosteric heat of adsorption, q_{st}, is obtained from

$$(\partial \ln p/\partial T)_\Gamma = q_{st}/RT^2 \qquad (6)$$

When a gas is adsorbed on a solid surface, it gives rise to a spreading pressure, π, which is defined (21) as

$$\pi = \gamma_s - \gamma_{sv} \tag{7}$$

where γ_s and γ_{sv} are the surface free energies of the solid at the solid-vacuum interface and at the solid-vapour interface. Spreading pressure may be calculated from the integrated form of the Gibbs' adsorption equation:

$$\pi = RT/p^o s \int (q/p)dp \tag{8}$$

where R is the gas constant, T is the temperature, q is the amount adsorbed, p^o is the saturated vapour pressure of the probe in units of $mol \cdot g^{-1}$, and s is the specific surface area of adsorbent. The spreading pressure may be obtained by graphical integration of the area under the curve of $q/(p/p^o)$ versus p/p^o. Spreading pressure at saturated vapour pressure, π^o, is of prime importance, since it can be related to the London force contribution of the surface free energy of the solid.

Combining the Fowkes (22) equation for the interfacial tension, $\gamma_{s\ell}$, between two phases with Young's equation for the contact angle, θ, of a liquid, ℓ, and a solid, s, when only London forces operate across the interface, a relationship is obtained between the equilibrium contact angle, θ, and the various tensions:

$$\cos \theta = -1 + [2(\gamma_s^L \gamma_\ell^L)^{\frac{1}{2}}]/\gamma_\ell - \pi^o/\gamma_\ell \tag{9}$$

Since the n-alkanes spread on carbon fibers, the contact angle, θ, is zero. In the case of hydrocarbons, $\gamma_\ell^L = \gamma_\ell$ and Equation 9 becomes

$$\gamma_s^L = [(\pi^o + 2\gamma_\ell)^2]/4\gamma_\ell \tag{10}$$

Materials and Methods

The carbon fibers used in this study, supplied by the Union Carbide Co., were: 1) T-300 3k (PAN) untreated, unsized; and 2) P-55 2k (pitch), untreated, unsized. Fibers were cut into 2 to 3 mm lengths and packed in 1.2 m glass columns with inner diameters of $4.0 \cdot 10^{-3}$ m. The adsorption was studied on "as received" (AR) and "cleaned" (T) fibers.

Zero coverage.   In order to eliminate physically adsorbed species, fibers were cleaned by heating at 160°C in a N_2 (Linde, ultra high purity, with CO_2 content less than 1 ppm) atmosphere until constant retention volumes were obtained (100 to 120 h).   Using finite concentration IGC and n-alkanes as sorbates, the surface area of these fibers was determined to be 0.40 $m^2 \cdot g^{-1}$ and 0.59 $m^2 \cdot g^{-1}$ for T-300 and P-55, respectively.   The n-alkanes octane to tridecane (analytical grade) were obtained from Polyscience Corporation (Quantkit).   Retention data were measured with a Hewlett-

Packard 5711A gas chromatograph equipped with dual flame ionization detectors. The nitrogen carrier gas (Linde, ultra high
purity) was passed through Linde 4Å molecular sieve. Ultra high
purity nitrogen was used because of its CO_2 low content (< 1 ppm),
which is readily physically-adsorbed on the carbon fiber surfaces
(23,24). Flow rates were measured at the column outlet with a
calibrated soap-bubble flow meter, and were corrected for the
vapour pressure of the soap solution in the flow meter (25),
column temperature (26), and pressure drop along the column.
Column pressures at the outlet were atmospheric and were determined with a barometer. The pressure drops, measured by a digital
pressure gauge (Setra 361) connected to the column inlet, ranged
from 25 to 37 $kN \cdot m^{-2}$, depending on flow rate and column. Flow
rates (~23 to 26 $cm^3 \cdot min^{-1}$) were maintained constant to within ±
1% during the day. Column temperatures, throughout any one set of
the measurements, were held constant to ± 0.05°C with a circulating water bath (Lauda K4R).

Hydrocarbon vapours were injected directly into the column
with a Hamilton (1 µℓ) syringe. Liquid volumes, equivalent to
those injected, are approximately 10^{-8} cm^3. The dead volumes of
the columns were determined by injecting methane, which was not
retained by the carbon fibers, simultaneously with the n-alkanes.

Finite Concentration. The carbon fiber samples and the column
preparation were essentially the same as in the zero coverage
region. The upper limit of 160°C for cleaning the carbon fiber
surface was chosen because decomposition of the surface groups
(for example, chemically-adsorbed oxygen) (27) occurs at higher
temperatures. It is recognized that when dealing with microporous
carbons, it is hopeless to rely on an outgassing temperature of
160°C because temperatures of 250 to 300°C are required (28).
This would result in partial removal of the above-mentioned surface groups. However, in the case of the untreated fibers used in
this work, the number of surface groups is relatively small compared to surface treated fibers. Hence, a sample of the untreated
fibers was heat treated at 300°C in N_2 for 12 h (sample HT).
n-Nonane has been used for deliberate blocking of the pores in
microporous carbon (29-31), so a single injection of n-alkane
would probably render this cleaning useless. In order to prevent
this effect, the column was heated after each injection for 20 min
in N_2. The temperatures that were used for this purpose were
successively higher (100, 150, 200, 250 and 300°C) in different
experiments.

The carbon fiber surface areas were previously determined by
BET krypton adsorption to be 0.62 ± 0.01 $m^2 \cdot g^{-1}$ and 0.74 ± 0.01
$m^2 \cdot g^{-1}$ for T-300 and P-55, respectively. The molecular area of
krypton was taken as 0.195 nm^2. Prior to these measurements, the
fibers were degassed at 300°C for 15 h. The 'elution of a characteristic point' method of finite concentration IGC was used to
determine the isotherms for a series of n-alkanes. Approximately
15 to 20 injections were used for each isotherm. The hand-drawn
curve through the peak maxima was digitized for integration and
subsequent data handling.

Adsorption of n-nonane on both types of "as received" (AR) fibers was studied at 30 ± 0.05°C. In the case of T-300T fibers, the column temperature was maintained constant at 60 ± 0.05°C except for n-nonane adsorption, which was also studied at 50 and 70°C. For P-55T fibers, the column temperature was maintained constant at 75 ± 0.05°C except for n-nonane adsorption, which was also studied at 50, 60 and 70°C. The temperature was controlled with a circulating water bath (Lauda K4R).

All measurements on columns containing HT fibers were made at 70 ± 0.5°C in an air oven. n-Nonane was the only sorbate used in these experiments.

The n-alkanes nonane to undecane were obtained from Polyscience (Quantkit) and Aldrich (Gold Label). IGC apparatus used to monitor adsorption isotherms is described in the zero coverage section. Flow rates in the range 23 to 32 $cm^3 \cdot min^{-1}$ were measured at the outlet of the column with a calibrated soap-bubble flow meter and were corrected for the vapour pressure of water in the flow meter. The pressure drops across the column were in the range 27 to 37 $kN \cdot m^{-2}$. Isotherms for each sorbate were obtained from a series of injections (from $\sim 10^{-8}$ cm^3, for zero coverage, to $6 \cdot 10^{-3} \cdot cm^3$), made with Hamilton 1 and 10 µL syringes. The dead volumes were determined by injecting methane. The retention volumes for methane were independent of temperature, thus showing no significant adsorption had occurred.

Results and Discussion

Zero Coverage. The peaks at infinite dilution were slightly skewed (skew ratio ~ 0.8), with virtually no dependence of retention volume on injection size. Instead of the peak maximum method, retention volumes were measured by the method proposed by Conder and Young (32). To ensure that the adsorption of n-alkanes on carbon fibers was taking place under equilibrium conditions, the flow rate was varied in the range 20 to 32 $cm^3 \cdot min^{-1}$. The net retention volumes were essentially independent of flow rate.

The Henry's law constants varied linearly with the temperature. Heats of adsorption (Table I) were calculated from Equation 3. Standard surface free energies of adsorption (Table I) were calculated from Equation 2 using de Boer's standard state for spreading pressure.

Table I shows that the differential heat of adsorption of n-alkane on "as received" carbon fibers is low and closely approximates its heat of liquefaction. This indicates a low concentration of high energy sites on the "as received" fibers. The differential heat of adsorption on "cleaned" fibers, especially T-300, is greater than on "as received" fibers, suggesting that some of the high energy sites on the carbon fiber surfaces were occupied by physically adsorbed species. GC analysis of desorption products, collected in a liquid nitrogen trap, showed the presence of water and carbon dioxide.

The values of the thermodynamic functions vary linearly with the number of carbon atoms, as shown in Table I. The observed increments in q_d (4.9 ± 0.1 $kJ \cdot mol^{-1}$ for T-300AR; 6.5 ± 0.1 kJ·

mol^{-1} for T-300T; 5.1 ± 0.3 $kJ \cdot mol^{-1}$ for P-55AR, and 5.7 ± 0.1
$kJ \cdot mol^{-1}$ for P-55T carbon fibers) indicate a flat orientation of
the n-alkanes on the carbon fiber surface. Constant increments
were similarly obtained for ΔG_A^0 at 50°C (-2.58 ± 0.03 $kJ \cdot mol^{-1}$ for
T-300AR; -3.64 ± 0.01 $kJ \cdot mol^{-1}$ for T-300T; -2.60 ± 0.01 $kJ \cdot mol^{-1}$
for P-55AR, and -3.05 ± 0.07 $kJ \cdot mol^{-1}$ for P-55T). Although the
adsorption potential of a methylene group on a surface is usually
estimated by the heat of adsorption, the free energy of adsorption
may also reflect this quantity. This holds when the free energy
and enthalpy changes for a homologous series are linearly related.
This is the case for the adsorption of n-alkanes on carbon fibers,
and $\Delta G_A^{(CH_2)} / q_d^{(CH_2)}$ is 0.53 for T-300AR; 0.56 for T-300T; 0.51 for
P-55AR; and 0.54 for P-55T at 50°C. According to Belyakova and
co-workers (33), this linear relationship is characteristic of
nonspecific adsorption.

Assuming that the work of adhesion between a saturated hydrocarbon and a second phase is equal to the free energy of desorption per mole of CH_2 groups, Dorris and Gray (5) proposed the
equation for the estimation of the London component of the surface
free energy of the adsorbent:

$$-\Delta G_A^{(CH_2)} / N \cdot a_{(CH_2)} = 2 \left(\gamma_{(CH_2)} \cdot \gamma_s^L \right)^{\frac{1}{2}} \tag{11}$$

where $a_{(CH_2)}$ is 0.06 nm^2, the cross-sectional area of a methylene
group; $\gamma_{(CH_2)}$ is the surface tension of a surface comprised of
only CH_2 groups (an extrapolation of the surface tension of the
low molecular weight n-alkanes to infinite chain length (34)
yields a value of 34.7 $mN \cdot m^{-1}$ at 20°C); and N is Avogadro's constant. Using the incremental free energy of adsorption obtained
experimentally at 50°C, the London component of the surface free
energy of carbon fibers was calculated from Equation 11 to be 38.8
$mN \cdot m^{-1}$ for T-300AR; 76.8 $mN \cdot m^{-1}$ for T-300T; 39.4 $mN \cdot m^{-1}$ for
P-55AR, and 54.4 $mN \cdot m^{-1}$ for P-55T. The value of 0.06 nm^2 for the
cross-sectional area of a methylene group is most widely used, but
it is worth mentioning that other values have been suggested.
Studying the adsorption of hydrocarbons on graphitized carbon
black, Clint (35) obtained the increment per CH_2 of 0.055 nm^2.
According to Groszek's model (36), the surface area of a methylene
group is 0.052 nm^2, which is equal to the area of the hexagon in
the graphite basal plane. Inserting Clint's or Groszek's value in
Equation 11 results in an increase of the London component of 20
or 30%, respectively.

Values for "as received" fibers agree with the values for
surface tension of Thornel 300 carbon fibers obtained by the
Wilhelmy and the solidification front technique (3) (42.4 and 41.8
$mN \cdot m^{-1}$). The method proposed by Kaelble (1) yields values (1,2)
of ~27 $mN \cdot m^{-1}$ and ~33 $mN \cdot m^{-1}$ for "as received" high-strength and
high-modulus fibers, respectively. These values seem to be too
low for the surface composed of graphitic basal planes and prismatic edge surfaces. In fact, these values are lower than the
London component of most organic polymers. Corresponding data for
"cleaned" fibers are not available in the literature.

Even though carbon fiber surfaces usually contain small amounts of hetero-atoms (for example, O, N), they are composed primarily of graphitic basal planes and prismatic edge surfaces, and thus their surface properties should be close to those of graphites and graphitized carbon black.

The surface free energy of the prismatic edge surfaces is much higher than the surface free energy of basal planes because their formation necessitates the breaking of covalent C-C bonds. Such a high surface free energy is rapidly lowered by chemical and physical adsorption, and thus it is difficult to obtain clean prismatic surface.

The London component of the surface free energy of "as received" carbon fibers is similar for both types of fibers. The increase that followed the cleaning process shows that some of the high-energy sites on both types of fibers were occupied by physically adsorbed species. Dispersion forces are proportional to the square of the density of material. The densities of the carbon fibers are ~1.75 $g \cdot mL^{-1}$ for T-300 and ~2 $g \cdot mL^{-1}$ for P-55. The density of the graphites is higher (~2.2 $g \cdot mL^{-1}$) and this may be responsible for the higher values of the London component. The degree of order in the P-55 fibers (high-modulus) is greater than in T-300 (high-strength) because of the higher graphitization temperature. The P-55 fibrils are well aligned, and the fiber surface has more basal planes and fewer edges than the surface of the T-300 fibers (37). The dispersion contribution to the surface free energy of the prismatic edge surfaces is higher than that of the basal planes. This difference explains the higher value of the London component for T-300 fibers, even though they have lower density than P-55.

Recently, Schultz et al. (38), presented an IGC study of carbon fiber and epoxy matrix surfaces. Their fiber conditioning treatment (105°C in helium for 20 h) was relatively mild compared to that used here for "cleaned" fibers; their results for the non-polar component of the surface free energy (γ_s^L = 50 ± 4 mN·m^{-1}) lies between the values reported here for "as received" and "cleaned" fibers. They also extended the method to estimate the polar component of the surface free energy from gas chromatographic measurements with polar vapours.

Finite Concentration. The adsorption isotherms for n-nonane, n-decane, and n-undecane on T-300T at 60°C are shown in Figure 1. Similar isotherm shapes were obtained for the same sorbates on P-55T fibers. The reproducibility of the isotherms for a second column filled with the same fibers was found to be ~2%.

Adsorption at a given p/p^o increased with decreasing carbon number. Assuming that all n-alkanes have the same surface orientation, this trend is expected. Interaction with n-alkanes occurring through dispersion forces and the amount adsorbed at each relative pressure is highly dependent on molecular size. The shape of the isotherms indicate that they belong to type II of the Brunauer-Deming-Deming-Teller classification (39). The experimental isotherms were interpreted according to the BET approach (40). The linear form of the BET isotherm is

Table I. Differential Heats and Standard Free Energies of Adsorption at Zero Coverage of n-Alkanes on Carbon Fibers at 50°C

Adsorbent	q_d (kJ mol⁻)					
	n-C	n-C	n-C	n-C	n-C	n-C
T-300AR		40.2	45.3	50.2	55.1	59.9
T-300T	51.1	57.6	64.2			
P-55AR		41.9	47.0	51.9	56.6	61.8
P-55T		49.6	55.4	61.1	66.8	72.5
	$- H_L$ (kJ mol⁻)*					
	39.8	44.6	49.4	54.2	58.9	63.7
	$- G_A^O$ (kJ mol⁻)					
T-300AR		20.8	23.4	26.0	28.6	31.1
T-300T	23.9	27.6	31.2			
P-55AR		20.7	23.4	26.0	28.6	31.2
P-55T		22.6	25.6	28.6	31.6	34.8

* Heat of liquefaction.

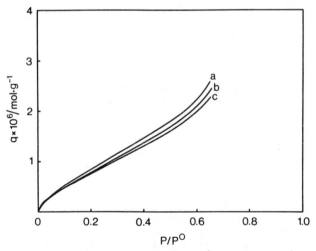

Figure 1. Adsorption isotherms for: a) n-nonane; b) n-decane; and c) n-undecane on T-300T carbon fibers at 60°C.

$$(p/p^o)/q(1-p/p^o) = 1/q_mC + [(C-1)/q_mC] \ p/p^o \qquad (12)$$

where q_m is the amount of probe adsorbed at monolayer coverage and C is the constant related to the heat of adsorption. If $(p/p^o)/q(1-p/p^o)$ is plotted against p/p^o, slope and intercept allow the estimation of q_m and C. BET plots are usually linear from 0.05 to 0.35 relative pressure. The location of the linear portion depends on the value of C and therefore on the heat of adsorption.

The major theoretical criticisms of BET are concerned with the assumptions that the surface is energetically uniform and that adsorbate-adsorbate interactions are negligible. In case of carbon fibers, the first assumption does not apply. Edge effects, due to the finite size of carbon layers and the presence of hetero-elements on the surface, are sources of energetic heterogeneity. The second assumption, that adsorbed molecules do not interact energetically, is also unrealistic. In spite of these inadequacies, the BET theory is useful in a qualitative sense and remains the most widely used method for surface area measurements.

BET plots in the range $0.1 < p/p^o < 0.03$ gave an excellent straight line for each hydrocarbon (correlation coefficients were in all cases > 0.988, while standard deviations of the slope and intercept were < 3%). Adsorption isotherms for n-nonane on T-300AR and P-55AR fibers yielded the C values of 2.7 and 2.8, respectively. Corresponding monolayer coverages were $1.15 \cdot 10^{-6}$ mol$\cdot$g^{-1} and $1.17 \cdot 10^{-6}$ mol$\cdot$g^{-1}. From the monolayer coverage, q_m, the specific surface area of adsorbent, s, can be calculated according to Equation 13, provided that the area occupied by each molecule, a_m, is known:

$$s = a_m \cdot N \cdot q_m \qquad (13)$$

where N is Avogadro's constant. From studies of the preferential adsorption of long chain n-alkanes from n-heptane solution onto Graphon and graphite, Groszek (14) suggests that, after slight compression of the n-alkanes, there is a remarkable fit between the hydrogen atoms attached to one side of the zig-zag carbon chains of n-alkanes and the centres of the hexagons formed by the carbon atoms in the basal plane. The compression needed to produce a perfect lattice fit requires the expenditure of energy. According to the calculations of Kiselev et al. (42) the interaction energy for adsorption with CH_2 groups located in the centres of the hexagons is ~25% higher than for adsorption on other sites. The extra energy gained by lattice matching is of the correct order of magnitude to compensate for the compression. Using Groszek's molecular areas for hydrocarbons on graphites, the surface areas of carbon fibers were calculated to be 0.40 m$^2 \cdot$g^{-1} and 0.59 m$^2 \cdot$g^{-1} for T-300T and P-55T, respectively. These values are smaller than those measured with krypton (0.62 m$^2 \cdot$g^{-1} and 0.74 m$^2 \cdot$g^{-1}), the reason probably being the conditioning of the samples, which was not the same in both cases.

Similar conditioning treatments for the HT fibers were used prior to IGC and krypton adsorption measurements. Adsorption isotherms for n-nonane on P-55HT at 70°C are shown in Figure 2. Corresponding BET parameters and the surface areas, calculated using Groszek's molecular area for n-nonane, are given in Table II.

The upward displacement of the adsorption isotherms after heat treatment, and the increase in q_m and C values may indicate the exposure of micropores after removal of adsorbed contaminants. The surface areas of these samples (0.61 $m^2 \cdot g^{-1}$ for T-300HT and 0.74 $m^2 \cdot g^{-1}$ for P-55HT) are in excellent agreement with the values from krypton adsorption.

The variation in isosteric heat of adsorption for n-nonane on T-300T fibers with surface coverage is presented in Figure 3. Also included are the values for q_{st} at $\Gamma = 0$, obtained by adding RT to the differential heat of adsorption, q_d, at zero coverage, the monolayer capacity for n-nonane at 50°C and the enthalpy of liquefaction for n-nonane at 50°C, $-\Delta H_L$.

A relatively high initial isosteric heat that decreases with increasing surface coverage is characteristic for heterogeneous surfaces. High energy sites are the first to interact with the sorbate molecules. As they become occupied, the adsorption takes place on other sites and the heat of adsorption slowly decreases. Both T-300T and P-55T fibers show such behaviour. Finally, at the monolayer coverage, as a result of increasing liquid-like character of the adsorbed n-nonane, q_{st} approaches $-\Delta H_L$. Similar results with graphite, carbon black, and graphitized carbon black have been obtained for the heat of adsorption of argon (43), krypton (44), and a number of hydrocarbons (45-48). The fact that the initial decrease for T-300T is spread over a higher surface coverage range than for P-55T is probably due to higher concentration of prismatic edge surfaces on these fibers. Such sites represent a major source of carbon fiber surface heterogeneity. The effect of lateral interactions that should lead to an increase in q_{st} at the point of monolayer coverage (49) is barely noticeable in both cases.

The London components of the surface free energy for T-300T, P-55T, T-300HT, and P-55HT fibers are listed in Table III. Values of 57.5 $mN \cdot m^{-1}$ and 62.8 $mN \cdot m^{-1}$ for P-55T and T-300T fibers, respectively from Table III, should be compared with values of 54.4 $mN \cdot m^{-1}$ and 76.8 $mN \cdot m^{-1}$ that were obtained for the same samples at zero surface coverage. In zero coverage calculations (Equation 11), the surface area of a methylene group was taken as 0.06 nm^2, while for finite concentration results (Equation 10 and Table III) it was taken as ~0.052 nm^2, according to Groszek's model. That means that the zero coverage values should be increased by ~30% prior to comparison. Thus, zero coverage values for the London components are ~100 $mN \cdot m^{-1}$ for T-300T and ~70 $mN \cdot m^{-1}$ for P-55T fibers. In both cases, zero coverage values are significantly greater than the finite coverage values. This can be explained by the heterogeneous character of the carbon fiber surfaces. Since the high energy sites are the first to interact with adsorbate molecules, zero coverage measurements give rise to the London

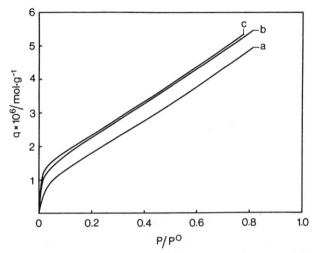

Figure 2. Adsorption isotherms for n-nonane on P-55HT fibers at 70°C: a) column heated at 70°C prior to each injection; b) at 150°C; and c) at 200°C.

Table II. BET Parameters for n-Nonane Adsorption on (HT) Carbon Fibers at 70°C and the Corresponding Specific Surface Areas*

Adsorbent	$q_m \cdot 10^6 (mol \cdot g^{-1})$	°C	$s(m^2 \cdot g^{-1})$
T-300HT-70**	1.50 ± 0.06	10.4 ± 0.4	0.53 ± 0.02
T-300HT-150**	1.54 ± 0.05	11.8 ± 0.3	0.54 ± 0.02
T-300HT-200**	1.69 ± 0.03	14.2 ± 0.3	0.59 ± 0.01
T-300HT-250**	1.75 ± 0.05	17.9 ± 0.6	0.61 ± 0.02
T-300HT-300**	1.74 ± 0.05	19.9 ± 0.6	0.61 ± 0.02
P-55HT-70**	1.87 ± 0.07	12.9 ± 0.4	0.65 ± 0.03
P-55HT-150**	2.14 ± 0.09	21.2 ± 0.8	0.74 ± 0.02
P-55HT-200**	2.15 ± 0.09	24.4 ± 0.9	0.74 ± 0.02

* Calculated using Groszek's value for a_m of n-nonane (0.575 nm^2).

** Denotes temperature at which column was heated prior to each injection (see Materials and Methods section).

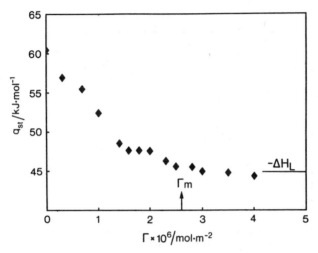

Figure 3.   The isosteric heat of adsorption of n-nonane as a function of the carbon fiber (T-300T) surface coverage at 50°C. ΔH_L is the enthalpy of liquefaction. Γ_m represents the mono-layer coverage.

Table III.   Equilibrium Spreading Pressures on n-Nonane on Carbon Fibers and the London Component of the Carbon Fiber Surface Free Energy at 70°C

Adsorbent	π^o (mN·m^{-1})	γ_s^L (mN·m^{-1})
T-300T*	30.9	62.8
T-300HT-70**	38.4	76.9
T-300HT-150**	39.0	78.2
T-300HT-200**	40.5	81.2
T-300HT-250**	41.6	83.6
T-300HT-300**	42.8	86.2
P-55T*	27.8	57.5
P-55HT-70**	34.2	68.5
P-55HT-150**	41.8	83.9
P-55HT-200**	42.9	86.4

*   Measurements at 50°C.
** Denotes temperature at which the column was heated prior to each injection (see Materials and Methods section).

component of such sites only. However, the finite coverage method
gives an overall value related to the average of all sites.
    It is interesting that the London components of "as received"
fibers calculated from spreading pressures of n-nonane (51 $mN \cdot m^{-1}$
and 52 $mN \cdot m^{-1}$ for T-300AR and P-55AR, respectively) are in good
agreement with the corresponding values from zero coverage
measurements (the values of 38.8 $mN \cdot m^{-1}$ for T-300AR and 39.4
$mN \cdot m^{-1}$ for P-55AR fibers have to be increased ~30% to bring them
to the same area basis for comparison). High energy sites on "as
received" fibers are occupied by physically adsorbed species (H_2O,
CO_2) and as a result, the heterogeneous character of their surface
is masked. Thus, zero coverage and finite coverage measurements
yield similar values for γ_s^L.
    The higher values of γ_s^L for T-300HT and P-55HT on heat-
treated fibers may be attributed to the enhanced dispersion force
field that the n-alkanes experience when adsorbed in micropores.
Gradual increases in treatment temperature cause gradual increases
in the London component, indicating that the n-alkanes are getting
deeper into the pores. Unlike the zero coverage values, London
components of spreading pressures are similar for T-300 and P-55
fibers. The higher concentration of prismatic edge sites on T-300
fibers, that was proposed to explain higher value of the London
component at zero coverage, is opposed by the effect of density.
The densities of the carbon fibers are ~1.75 $g \cdot mL^{-1}$ for T-300 and
~2 $g \cdot mL^{-1}$ for P-55. Dispersion forces are proportional to the
square of the density of material. Hence, the higher density of
P-55 fibers (effective at all coverages), compensates for the
contribution of high-energy sites (effective only at low coverages) on T-300 fibers.

Conclusion

IGC has been shown to be an effective tool for studying the surface properties of carbon fibers. Results for the heat of adsorption indicate that "as received" carbon fibers are low energy surfaces for the adsorption of n-alkanes. The London component supports this. Desorption of physically adsorbed species (CO_2, H_2O)
results in a significant increase in the heat of adsorption and
the London component of the surface free energy for both high-
strength and high-modulus fibers. This shows that the high energy
sites were occupied by physically adsorbed gases. The London
component of the surface free energy is higher for "cleaned" high-
strength fibers (76.8 $mN \cdot m^{-1}$) than for high-modulus fibers (54.4
$mN \cdot m^{-1}$). The reason for this is probably a higher content of high
energy prismatic edge surfaces on the former. Type II adsorption
isotherms for hydrocarbons on high-modulus (P-55) and high-
strength (T-300) carbon fibers were calculated from the IGC peaks
in the finite concentration region. The isotherms are well described by the BET equation and monolayer coverages were easily
determined.
    Increases in q_m and C values with the increase in temperature
of column conditioning indicates the presence of micropores.
Carbon fiber surface areas, calculated with Groszek's values for

the molecular areas of hydrocarbons on graphites, are in excellent agreement with those calculated from krypton adsorption. Variation of thermodynamic properties with surface coverage on both types of fibers indicates the presence of high-energy surface sites. From spreading pressures, the London components of the carbon fibers, surface free energy at $50°C$ are estimated to be 57.5 mN·m^{-1} and 62.8 mN·m^{-1} for P-55T and T-300T, respectively. These values are lower than the corresponding zero coverage estimates (~70 mN·m^{-1} for P-55T and ~100 mN·m^{-1} for T-300T). This is explained by the heterogeneous character of the carbon fiber surfaces. High-energy sites interact first with the adsorbate molecules, and zero coverage measurements reflect the London component of such sites. However, the finite coverage method gives an overall value related to the average of all sites. High-energy sites on AR fibers are occupied by the physically adsorbed species (H_2O, CO_2), which mask the surface heterogeneity. In this case, zero and finite coverage measurements yield similar values for the London component.

Legend of Symbols

$a_{(CH_2)}$	Surface area occupied by a CH_2 segment of an n-alkane (m^2).
a_m	Surface area occupied by adsorbate molecule (m^2).
A	Surface area stationary phase in column (m^2).
c	Mobile phase probe concentration (mol/mL).
C	Constant in BET equation.
K_s	Surface partition coefficient (mL/m^2).
N	Avogadro's constant.
p	Vapour pressure of probe (kPa).
p^o	Saturated vapour pressure of probe (kPa).
$P_{s,g}$	Vapour pressure in standard gaseous state $(101\ kPa)$.
q	Surface excess of probe per unit weight of stationary phase (mol/g).
q_d	Differential heat of adsorption (kJ/mol).
$q_d^{(CH_2)}$	Increment in q_d per CH_2 for a series of n-alkanes.
q_m	Value of q corresponding to monolayer average (mol/g).
q_{st}	Isosteric heat of adsorption (kJ/mol).
R	Gas constant.
s	Specific surface area of adsorbent (m^2/g).
T	Column temperature (K).
V_N	Net chromatographic retention volume (mL).
w	Weight of adsorbent stationary phase in column (g).
$\Delta G_A^{(CH_2)}$	Free energy of adsorption of a CH_2 segment of n-alkane.
ΔG_A^o	Standard free energy of adsorption (kJ/mol).
ΔH_L	Enthalpy of liquefaction of the probe vapour (kJ/mol).
γ_ℓ	Surface tension or surface free energy of liquid (mN/m).
γ_s	Surface free energy of solid in vacuo (mN/m).
γ_ℓ^L	London component of surface free energy of liquid (mN/m).

γ_s^L — London component of surface free energy of solid (mN/m).

$\gamma_{s\ell}$ — Solid-liquid interfacial tension (mN/m).

γ_{sv} — Surface free energy of solid-vapour interface (mN/m).

Γ — Surface excess of probe per unit surface area of stationary phase (mol/m²).

π — Spreading pressure of probe on surface (mN/m).

π^o — Spreading pressure of probe at saturated vapour pressure (mN/m).

π_s — Spreading pressure of de Boer standard adsorbed state (0.338 mN/m).

Θ — Solid-liquid contact angle.

Literature Cited

1. Kaelble, D.H.; Dynes, P.J.; Cirlin, E.H. J. Adhesion 1974, 6, 23.
2. Hammer, G.E.; Drzal, L.T. Applications of Surf. Sci. 1980, 4, 340.
3. Li, S.K.; Smith, R.P.; Neumann, A.W. J. Adhesion 1984, 17, 105.
4. Schultz, J.; Cazeneuve, C.; Shanahan, M.E.R.; Donnet, J.B. J. Adhesion 1981, 12, 221.
5. Dorris, G.M.; Gray, D.G. J. Colloid Interface Sci. 1980, 77, 353.
6. Katz, S.; Gray, D.G. J. Colloid Interface Sci. 1981, 82, 318.
7. Anhang, J.; Gray, D.G. J. Appl. Polym. Sci. 1982, 27, 71.
8. Gozdz, A.S.; Weigmann, H.D. J. Appl. Polym. Sci. 1984, 29, 3965.
9. Domingo-Garcia, M.; Fernandez-Morales, I.; Lopez-Garzon, F.J.; Moreno-Castilla, C. J. Chromatography 1984, 294, 41.
10. Vukov, A.J.; Gray, D.G. Langmuir. 1988, 4, 743.
11. de Boer, J.H. The Dynamical Character of Adsorption; Clarendon Press: Oxford, 1953; Chapter VI.
12. Kiselev, A.V.; Yashin, Y.I. Gas Adsorption Chromatography; Plenum: New York, 1969; Chapter IV.
13. Gray, D.G.; Guillet, J.E. Macromolecules 1972, 5, 316.
14. Mohlin, U.B.; Gray, D.G. J. Colloid Interface Sci. 1974, 47, 747.
15. Tremaine, P.R.; Gray, D.G. J. Chem. Soc. Faraday Trans. I, 1975, 71, 2170.
16. Dorris, G.M.; Gray, D.G. J. Colloid Interface Sci. 1979, 71, 93.
17. Katz, S.; Gray, D.G. J. Colloid Interface Sci. 1981, 82, 326.
18. Saint Flour, C.; Papirer, E. Ind. Eng. Prod. Res. Dev. 1982, 21, 337.
19. Domingo-Garcia, M.; Lopez-Garzon, F.J.; Lopez-Garzon, R.; Moreno-Castilla, C. J. Chromatography 1985, 324, 19.
20. James, A.T.; Martin, A.J.P. Biochem. J. 1952, 50, 679.

21. Aveyard, R.; Haydon, D.A. An Introduction to the Principles
     of Surface Chemistry; Cambridge University Press: London,
     1973; p.150.
22. Fowkes, F.M. In Chemistry and Physics of Interfaces; Ross,
     S., Ed.; American Chemical Society Publications: Washington,
     D.C., 1965; p.1.
23. Drzal, L.T. Carbon 1977, 15, 129.
24. Drzal, L.T.; Meschner, J.A.; Hall, D. Carbon 1979, 17, 375.
25. Karger, B.L.; Snyder, L.R.; Horvath, C. An Introduction to
     Separation Science; Wiley-Interscience: New York, 1973;
     p. 215.
26. Littlewood, A.B. Gas Chromatography; Academic Press: New
     York, 1970, 2nd ed., p.31.
27. Donnet, J.B.; Papirer, E.; Couderc, P. Bull. Soc. Chim. Fr.
     1968, 3, 929.
28. Sing, K.S.W. Carbon 1987, 25, 151.
29. Gregg, S.J.; Langford, J.F. Trans. Faraday Soc. 1969, 65,
     1394.
30. Rodriguez-Reinoso, F.; Martin-Martinez, J.M.; Molina-Sabio,
     M.; Torregrosa, R.; Garrido-Segovia, J. J. Colloid Interface
     Sci. 1985, 106, 315.
31. Ali, S.; McEnaney, B. J. Colloid Interface Sci. 1985, 107,
     355.
32. Conder, J.R.; Young, C.L. Physicochemical Measurements by Gas
     Chromatography; Wiley: Chicester, 1979; p. 82.
33. Belyakova, L.D.; Kiselev, A.V.; Kovaleva, N.V. Russ. J. Phys.
     Chem. 1966, 40, 811.
34. Aveyard, R. J. Colloid Interface Sci. 1975, 52, 621.
35. Clint, J.H. Trans Faraday Soc. 1972, 68, 2239.
36. Groszek, A.J. Proc. Roy. Soc. 1970, A 314, 473.
37. Guinon, M.; Oberlin, A. Proc. of Carbon '86, 4th Int. Carbon
     Conf. 1986, Deutsch. Ker. Ges., Baden-Baden, p.614.
38. Schultz, J.; Lavielle, L.; Martin, C. J. Chim. Phys. 1987,
     84, 231.
39. Brunauer, S.; Deming, L.S.; Deming, W.S.; Teller, E. J. Amer.
     Chem. Soc. 1940, 62, 1723.
40. Brunauer, S.; Emmet, P.H.; Teller, E. J. Amer. Chem. Soc.
     1938, 60, 309.
41. Groszek, A.J. Proc. Roy. Soc. 1970, A 314, 473.
42. Avgul, N.N.; Isirikyan, A.A.; Kiselev, A.V.; Lygina, I.A.;
     Poshkus, D.P. Izvest. Akad. Nauk. S.S.S.R., Otdel. Khim.
     Nauk. 1957, 1314.
43. Grillet, Y.; Rouquerol, F.; Rouquerol, J. J. Colloid Inter-
     face Sci. 1979, 70, 239.
44. Putnam, F.A.; Fort, T. Jr. J. Phys. Chem. 1975, 79, 459.
45. Isirikyan, A.A.; Kiselev, A.V. J. Phys. Chem. 1962, 66, 210.
46. Taylor, G.L.; Atkins, J H. J. Phys. Chem. 1966, 70, 1678.
47. Dollimore, D.; Heal, G.R.; Martin, D.R. J. Chem. Soc.,
     Faraday Trans. I. 1972, 68, 832.
48. Crescentini, G.; Mangani, F.; Mastrogiacomo, A.R.; Palma, P.
     J. Chromatography 1987, 392, 83.
49. Gregg, S.J.; Sing, K.S.W. Adsorption, Surface Area and Poros-
     ity 2nd ed., Academic Press: London, 1982, Chapter 2.

RECEIVED November 28, 1988

Chapter 14

Interfacial Properties of Carbon Fiber–Epoxy Matrix Composites

Jacques Schultz[1] and Lisette Lavielle[2]

[1]Centre de Recherches sur la Physico-Chimie des Surfaces Solides, Centre National de la Recherche Scientifique, 24 avenue du Président Kennedy, 68200 Mulhouse, France
[2]Laboratoire de Recherches sur la Physico-Chimie des Interfaces de l'Ecole Nationale Supérieure de Chimie de Mulhouse, 3 rue Alfred Werner, 68093 Mulhouse Cedex, France

The performance of a composite material depends strongly on the quality of the fibre-matrix interface. The interactions developed at the carbon fibre-epoxy matrix interface were studied using the acid/base or acceptor/donor concept. The surface characteristics of the reinforcing fibre and the polymer were determined using a tensiometric method and the inverse gas chromatography technique at infinite dilution. Following the approach of Gutmann, acid/base surface characteristics were obtained, allowing the interactions at the interface to be described by a specific interaction parameter. It was shown that the shear strength of the interface, that is, the capacity of the interface to transfer stress from the matrix to the fibre, as measured by a fragmentation test, is strongly correlated to this specific interaction parameter, demonstrating the great importance of acid/base interactions in fibre-matrix adhesion.

The adhesion between reinforcing fibres and the polymer matrix is one of the important parameters governing the performance of a composite material. It is generally recognized that the interface or interphase is the third constituent of a composite material. However, high adhesive strength at the interface does not necessarily lead to optimum properties of the composite. For instance, good impact strength would require the formation of reversible physical bonds rather than high energy chemical links. In order to understand and predict the mechanical behaviour of the composite, it is necessary to gain better knowledge of the nature and level of interactions likely to be exchanged at the interface.

In the case of physical bonds (London dispersion, Keesom orientation, and Debye induction forces), the energy of interaction or reversible energy of adhesion can be directly calculated from the surface free energies of the solids in contact.

For years, it was assumed that the surface energy γ_S of a solid is the sum of two terms: a dispersive component γ_S^D and a non-dispersive or polar component γ_S^P. However, recently, it appears that the non-dispersive term of interaction could be better described using the concept of electron acceptor/donor or Lewis acid/base characteristics of the solids.

0097–6156/89/0391–0185$06.00/0
© 1989 American Chemical Society

Extending work done previously (1 - 3), the purpose of this paper is to examine how these characteristics could be determined using inverse gas chromatography (IGC) and to what extent these acid/base interactions are relevant to the description of the fibre-matrix interface.

Materials

Three high-strength PAN (polyacrylonitrile)-based carbon fibres (supplied ·by Elf Aquitaine France), corresponding to three different stages of manufacturing, were used in this study:
- the untreated fibre 1;
- the untreated fibre having received a proprietary surface treatment, designated as oxidized fibre 2; and
- the oxidized fibre having received a supplementary sizing treatment, denoted as coated fibre 3.
    Two other fibres were also studied, although less extensively:
- a fibre having been oxidized electrolytically on a pilot apparatus, denoted as oxidized fibre 2'; and
- another commercial coated high strength PAN-based fibre, denoted as coated fibre 4.
    The matrices were two DGEBA (diglycidyl ether of bisphenol A) epoxy resins. The hardeners were either 35 parts by weight of diamino diphenyl sulfone (resin I) or 55 parts of polyamino amide (resin II). The curing conditions were 3 h at 130°C followed by 3 h at 180°C for resin I and 24 h at 40°C followed by 6 h at 100°C for resin II.

Methods

Inverse Gas Chromatography. Chromatographic measurements at infinite dilution were carried out with an Intersmat IGC 120 DLF equipped with a flame ionization detector of high sensitivity. The chromatograph was coupled with a Shimadzu integrator, allowing for automatic analysis of the first moment of the elution peak to be made. The fibers or resin particles (obtained by grinding in liquid nitrogen) were packed into stainless steel columns of 0.6 m length and 4.4 mm internal diameter.
    The carrier gas was helium. The amount of probe injected corresponds to 10^{-4} to 10^{-3} ppm, thus ensuring practically infinite dilution or zero surface coverage. These conditions allowed the study of only probe-adsorbant interactions, the adsorbed molecules being sufficiently far apart to neglect their mutual interaction. These optimum working conditions produced symmetrical peaks that followed the laws of linearity and ideality required for their interpretation.
    The fundamental quantity of inverse gas chromatography, V_N, the net retention volume, was determined using:
$$V_N = j \, D \, (t_R - t_O),$$

where t_R is the retention time of the given probe, t_O is the zero retention reference time measured with a practically non-adsorbing probe such as methane, D is the corrected flow rate and j is a correction factor taking into account gas compressibility.
    Before experimentations, the IGC columns were conditioned in helium at 105°C for 20 h.

Wetting. In the case of low surface energy solids, such as polymers, the dispersive and polar components of surface energy are easily determined through contact angle measurements. Drops of various liquids of known surface energy are deposited onto the solid surface and the contact angles are measured using a Rame-Hart apparatus.

Dispersive and polar components of the surface energy of the solids are deduced from the classical $\cos \theta = f [(\gamma_L^D)^{1/2}/\gamma_L]$ (4), where γ_L is the surface energy of the wetting liquid and γ_L^D its dispersive component.

In the case of high surface energy solids, virtually all liquids spread spontaneously on the surface. Therefore, a two phase liquid method was used (4, 5). In this method, a drop of a polar liquid, L, is placed on the solid surface, S, the surrounding medium being a non-polar hydrocarbon H, immiscible with the contact liquid (Figure 1).

The analysis of the system leads to a general relationship between the measured contact angle $\theta_{SL/H}$ and the different surface and interfacial energies of the three constituents of the system :

$$\underbrace{\gamma_L - \gamma_H + \gamma_{HL} \cos \theta_{SL/H}}_{Y} = 2(\gamma_S^D)^{1/2} \underbrace{[(\gamma_L^D)^{1/2} - (\gamma_H)^{1/2}] + 2(\gamma_S^P \gamma_L^P)^{1/2}}_{X} \quad (1)$$

where γ_H is the surface energy of the hydrocarbon and γ_{HL} the interfacial energy between the hydrocarbon and the polar liquid.

By plotting the quantity Y versus the quantity X, a straight line is obtained, allowing the determination of γ_S^D from the slope and γ_S^P from the intersection at the origin.

With carbon fibres, such direct measurements would be extremely difficult since cylindrical solids with diameters of 7 to 10 μm are being considered.

The method used is tensiometric (6) and depends on the fact that a fibre immersed in a two phase liquid system results in the formation of a meniscus, leading to an apparent increase in the weight of the fibre (Figure 2).

Experimentally, the single fibre is attached to the arm of an electro-balance and immersed first in the hydrocarbon alone, and then in both hydrocarbon and polar liquid phases. Weight increases at each stage are measured. Static and dynamic experiments, in immersion and emersion, have been conducted.

Because of the small diameter of the fibre, the buoyancy force can be neglected. Thus, at the first stage of immersion, the apparent weight increase, is due to the meniscus of hydrocarbon touching the fibre surface, as given by:

$$F_{HA} = C \gamma_H \cos \theta_{SH/A} \quad (2)$$

Because of the low surface energy of the alkane, the contact angle $\theta_{SH/A}$ is zero and the circumference C of the fibre is readily evaluated. At the second stage of immersion, the apparent weight increase is due to the meniscii alkane/polar liquid (water or formamide), as given by:

$$F = F_{HA} + F_{HL} = C \gamma_H + C\gamma_{HL} \cos \theta_{SL/H} \quad (3)$$

Since C, γ_H and γ_{HL} are known, the contact angle $\theta_{SL/H}$ can be calculated. The same analysis as that used for flat surfaces may then be applied to carbon fibres.

Fragmentation Test. The problem of determining fibre-matrix adhesion has received considerable attention. The analysis using the fragmentation of a model system is considered to constitute the best solution.

An elementary carbon fibre is embedded in an epoxy matrix and this model composite is submitted to an uniaxial tensile load in the direction of fibre orientation.

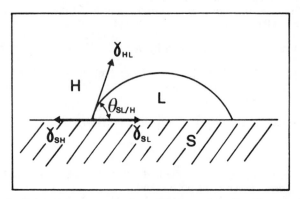

Figure 1: Schematic representation of the two-phase liquid method.

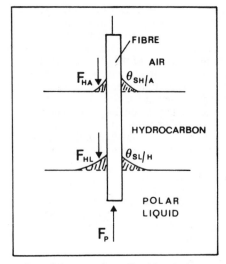

Figure 2: Representation of the two-phase liquid technique applied to a carbon fibre.

This load is transmitted through the interface from the matrix to the fibre and, as originally described by Kelly and Tyson (7), the fibre breaks into small fragments until a limiting fragment size l_c is reached (Figure 3). Knowledge of l_c enables one to determine the shear resistance of the interface; that is, the capacity of the interface to transfer the stress from the matrix to the fiber. This shear resistance of the interface , τ, is therefore a direct measurement of fibre-matrix adhesion.

According to the Kelly-Tyson model (7) and Cox theory (8), both maximum shear strength τ_{max} at the extremities of the fragment and average shear strength τ can be determined knowing the critical length l_c of the fragments and the tensile strength of the carbon fibre at this critical length. Fragment length distribution was determined using an optical microscope equipped with a micrometer eyepiece. Critical length and tensile strength were obtained by means of Weibull statistics (9, 10).

Results

Determination of Surface Properties Using Inverse Gas Chromatography.

Theory (11). Simple thermodynamic considerations applied to inverse gas chromatography at infinite dilution lead to the following general relationship:

$$\Delta G_D^o = - \Delta G_A^o = RT \, Ln \left(\frac{V_N P_o}{S g \, \pi_o} \right) \tag{4}$$

where ΔG^o is the free enthalpy of desorption (or adsorption) of 1 mole of solute from a reference adsorption state, defined by the bidimensional spreading pressure π_o of the adsorbed film to a reference gas phase state, defined by the partial pressure P_o of the solute,

S, is the specific surface area of the substrate (in this case, fiber),

and   g, is the weight of fibre substrate in the column.

Two reference states are generally considered:
the reference state of Kemball and Rideal (12), where $P_o = 1.013 \cdot 10^5$ Pa and $\pi_o = 6.08 \cdot 10^{-5}$ N·m^{-1}; and the reference state of De Boer (13), where $P_o = 1.013 \cdot 10^5$ and $\pi_o = 3.38 \cdot 10^{-4}$ N·m^{-1}.

Therefore, ΔG^o can be written as
$$\Delta G^o = RT \, Ln \, V_N + K, \text{ where} \tag{5}$$
K is a constant for a given chromatographic column depending on the chosen reference states.

To a first approximation, ΔG^o is related to the energy of adhesion, W_A, between the probe molecule and the solid, per unit surface area of the solid by
$$\Delta G^o = N \, a \, W_A, \text{ where} \tag{6}$$
N is Avogadro's number and a is the surface area of the probe molecule. Combining Equations 5 and 6 leads to :
$$RT \, Ln \, V_N = N \, a \, W_A + Constant \tag{7}$$
where this constant, as mentioned previously, depends only on the reference states.

Dispersive component of surface energy. According to Fowkes (14), when only dispersion interactions are being exchanged, for example with n-alkanes probes, the energy of adhesion is given by

$$W_A = 2 \, (\gamma_S^D \, \gamma_L^D)^{1/2} \tag{8}$$

Therefore, Equation 7 can be written as

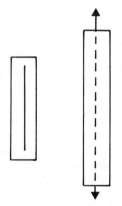

Figure 3: Schematic representation of the fragmentation technique.

$$RT \, Ln \, V_N = 2 N \, (\gamma_S^D)^{1/2} \, a \, (\gamma_L^D)^{1/2} + C^t \qquad (9)$$

or, in the case of n-alkanes (where $\gamma_L^D = \gamma_H$):

$$RT \, Ln \, V_N = 2 N \, (\gamma_S^D)^{1/2} \, a \, (\gamma_H)^{1/2} + C^t \qquad (10)$$

As shown, for the examples given in Figure 4, $RT \, Ln \, V_N$ is a linear function of the quantity $a(\gamma_H)^{1/2}$. The slope of the straight line leads to the γ_S^D values of the carbon fibres and matrices listed in Table I.

Table I. Dispersive Component of Surface Energy at 25°C (in mJ·m^{-2})

		IGC			
		Our Analysis γ_S^D	Gray's Analysis γ_S^D	Wetting Method* γ_S^D	γ_S^D
Fibres	Untreated 1	50 ± 4	48 ± 4	50 ± 8	7 ± 3
	Oxidized 2	49 ± 2	50 ± 4	48 ± 10	15 ± 4
	Oxidized 2'	59 ± 2	-	-	-
	Coated 3	36 ± 3	33 ± 3	34 ± 6	13 ± 4
	Coated 4	29 ± 3	31 ± 4	-	-
Matrices	Epoxy I	39 ± 3	41 ± 3	40 ± 3	4 ± 1
	Epoxy II	34 ± 7	-	37 ± 3	19 ± 2

*the γ_S^D values of the fibres were determined using the two/phase liquid technique whereas for the matrices the one-liquid techique was employed.

It must be stressed that in our analysis, the γ_L or γ_H value used corresponds to a liquid, although the molecules adsorbed at infinite dilution can hardly be compared to an adsorbed liquid film. Therefore, in order to check the validity of the approximation contained in this intrepetation, the results were compared with those obtained using either an analysis developed by Gray (15-17) or the wetting method.

Gray uses a method for the determination of γ_S^D that considers the increment

$$\Delta G_{(CH_2)}^o$$

per methylene group in the n-alkanes series with the general formula $C_n H_{2n+2}$. The increment

$$\Delta G_{(CH_2)}^o$$

as defined by

$$\Delta G_{(C_{n+1} H_{2n+4})}^o - \Delta G_{(C_n H_{2n+2})}^o$$

leads to

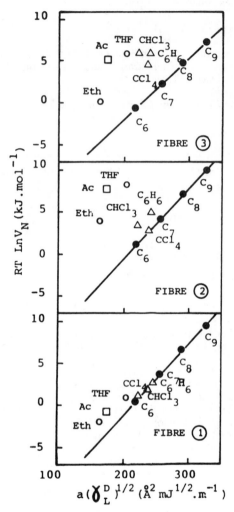

Figure 4: RT Ln V_N versus $a(\gamma_L^D)^{\frac{1}{2}}$ plot for the three carbon fibres. (Reproduced with permission from ref. 3. Copyright 1987 Gordon and Breach.)

$$\gamma_s^D = \frac{\left[RT \, Ln \, \dfrac{V_{N(C_{n+1} H_{2n+4})}}{V_{N(C_n H_{2n+2})}} \right]}{4N^2 \, a_{CH_2}^2 \, \gamma_{CH_2}} \qquad [11]$$

where a_{CH_2} is the surface area of a CH_2 group; that is, 6 Å^2, and γ_{CH_2} is the surface energy of a CH_2 group; that is, of a surface constituted of close packed CH_2 groups analogous to polyethylene and given by

$$\gamma_{CH_2} = 35.6 \, mJ.m^{-2} \quad with \quad \frac{d\gamma_{CH_2}}{dT} = 0.058$$

The values of γ_s^D obtained using Gray's method are also presented in Table I and agree satisfactorily with those obtained by our analysis.

The one-liquid method was used to determine the surface energy of the polymer matrix, whereas the two-phase liquid method was applied to measure the surface properties of the carbon fibres.

An example of a diagram obtained in the case of the two-phase liquid (hydrocarbons H and formamide F) method is shown in Figure 5. As pointed out in the Methods section, the slope of the straight line leads to the value of γ_s^D. The surface polarity or polar component of surface energy is determined from the intercept.

The values obtained by the tensiometric wetting method are also given in Table I.

It must be noted that the chromatographic method leads to values of γ_s^D that are more precise than the ones obtained by wetting, although they agree well. The untreated and oxidized carbon fibres have high values of γ_s^D, whereas the coated fibres exhibit somewhat lower values, close to the ones for the polymer matrices.

The value for the coated fibre is in agreement with the fact that the coating is usually made of either epoxy or polyester.

The value of γ_s^D for the virgin fibre is in agreement with other published values (18-20), but is lower than values obtained for graphite (21). One reason for this may be the mild conditions of the fibre treatment in the column (105°C for 20 h). However, it must be noted that these conditions can be considered relevant in view of the curing conditions applied to the composite.

Specific interactions. As demonstrated in the previous section, the tensiometric method also provides values of the surface polarity of the solids (Table I). The surface polarity of the untreated fibre is low whereas that of the oxidized or coated fibres is high. Furthermore, γ_s^P is different for the two matrices.

The purpose of this section is to describe the fibre and matrix surfaces in terms of electron acceptor or donor (acid/base) characteristics. According to this concept developed by Drago (22-23), Gutmann (24) and extended by Fowkes (25-27), strong interactions develop only between an acid and a base. Materials of the same characteristics, acids or bases even with high surface polarities, exchange nearly zero specific interaction. In this work, Gutmann's approach was adopted, rather than Drago's because it then becomes possible to consider a liquid or a solid to be both an acid and a base.

According to Gutmann (24), liquids can be characterized by donor or acceptor numbers. The donor number DN, defining the bacisity or electron-donor ability, is the molar enthalpy for the reaction of the electron-donor D with a reference acceptor $SbCl_5$. The acceptor number AN, characterizing the acidity or electron-acceptor ability is defined as the NMR chemical shift of ^{31}P contained in $(C_2H_5)_3$ PO when reacting with acceptor A.

In this study, several specific probes were chosen, exhibiting either a strong donor (base) trait or a strong acceptor (acid) trait, or both characteristics simultaneously (amphoteric).

Table II shows the main characteristics of some of the probes used in this work. The surface area a of the probe molecules was determined by injecting the probes onto neutral reference solids (PTFE, polyethylene, etc). The dispersive component γ_L^D was measured by the contact angle method on reference solids (4). The values of DN and AN were taken from tables published by Gutmann (24).

Table II. Characteristics of Some Probes

	a ($Å^2$)	γ_L^D ($mJ \cdot m^{-2}$)	DN	AN	Specific Characteristic
C_6H_{14}	51.5	18.4	-	-	
C_7H_{16}	57.0	20.3	-	-	Neutral
C_8H_{18}	62.8	21.3	-	-	
C_9H_{20}	68.9	22.7	-	-	
T H F	45	22.5	20.0	8.0	
Ether	47	15	19.2	3.9	Base
$CHCl_3$	44	25.0	0	23.1	
CCl_4	46	26.8	0	8.6	Acid
C_6H_6	46	26.7	0.1	8.2	
Acetone	42.5	16.5	17.0	12.5	
Ethyl - Acetate	48	19.6	17.1	9.3	Amphoteric

In order to determine quantitatively these specific interactions, consider, as a first approximation, that the specific interactions are simply added to the dispersive interactions defined previously. Therefore, the experimental point corresponding to a probe capable of specific interaction (that is, acid or base characteristics) always lies above the reference straight line of RT Ln V_N versus $a(\gamma_L^D)^{1/2}$ corresponding to the n-alkanes, as schematically illustrated in Figure 6. At a given value of $a(\gamma_L^D)^{1/2}$, the difference of ordinates between the point corresponding to the specific probe and the reference line leads to the value of the free enthalpy of desorption ΔG_{sp}^0 corresponding to specific acid - base interactions.

$$RT\,Ln\,\frac{V_N}{V_N^{ref}} = \Delta G_{sp}^o \qquad [12]$$

Such experiments were carried out on the carbon fibres and matrices at various temperatures. An illustration of the RT LnV_N versus $a(\gamma_L^D)^{1/2}$ diagrams obtained at 40°C for some carbon fibres is presented in Figure 4.

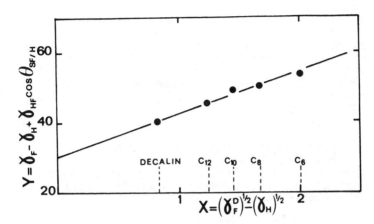

Figure 5: Determination of γ_S^D and γ_S^P of the coated fibre 3 by the wetting technique.

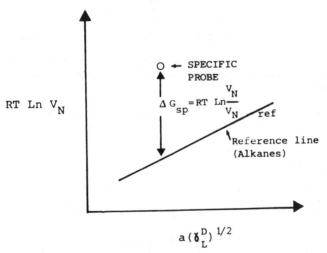

Figure 6: Schematic representation of a general $RT \ Ln \ V_N$ versus $a(\gamma_L^D)^{1/2}$ diagram.

It can be concluded, on a qualitative basis, that:
- the untreated fibre 1 has a moderate acid or acceptor characteristic and practically no base or donor characteristics;
- the oxidized fibre 2 has a strong acid characteristic and a rather low base characteristic; and
- the coated fibre 3 could be called amphoteric, since it exhibits a strong acid characteristic together with a high base characteristic.

In order to get at least a semi-quantitative approach to the acid-base surface properties of the solids, the enthalpy of desorption, ΔH_{sp}, corresponding to the specific interactions was determined by studying the variation of ΔG_{sp} with temperature T according to :

$$\Delta G_{sp} = \Delta H_{sp} - T \Delta S_{sp} \tag{13}$$

Following Papirer's approach (28), it was assumed that
$$\Delta H_{sp} = K_A \cdot DN + K_D \cdot AN \tag{14}$$
where DN and AN are Gutmann's numbers for the probes, and K_A and K_D are numbers describing the acid and base characteristics of the fibres or matrices.
Equation 14 can be written as

$$\frac{\Delta H_{sp}}{AN} = K_A \frac{DN}{AN} + K_D \tag{15}$$

Figure 7 shows that a plot of $\dfrac{\Delta H_{sp}}{AN}$ versus $\dfrac{DN}{AN}$ is actually linear.
Therefore, K_A and K_D can be determined from the slope and intercept at the origin of this straight line.

Table III gives the values of K_A and K_D calculated for the carbon fibres and the matrices. This quantitative approach leads to the same conclusions as those drawn from the qualitative examination.

Table III. Acid/Base (Acceptor/Donor) Characteristics (in Arbitrary Units)

		K_A	K_D
Fibres	Untreated 1	6.5 ± 0.5	1.5 ± 0.5
	Oxidized 2	10.0 ± 0.7	3.2 ± 0.2
	Oxidized 2'	9.7 ± 0.7	3.6 ± 0.3
	Coated 3	8.6 ± 0.7	13.0 ± 1.2
	Coated 4	9.1 ± 0.3	9.3 ± 0.3
Matrices	Epoxy I	7.6 ± 0.6	6.2 ± 0.5
	Epoxy II	10.1 ± 0.8	7.6 ± 0.6

Knowing the K_A and K_B values for the fibres and matrices and by analogy with Equation 14, it is possible to define a specific interaction parameter A, describing the acid/base interaction between the fibre (f) and the matrix (m) as

$$A = K_{A(f)} K_{D(m)} + K_{A(m)} K_{D(f)} \tag{16}$$

The calculated values of A are shown in Table IV. The specific interaction

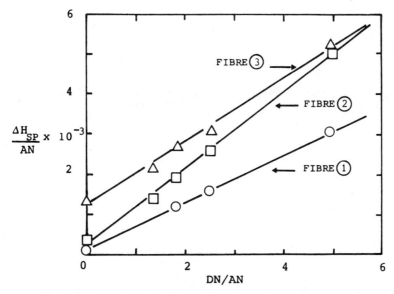

Figure 7: Determination of K_A and K_D values of the carbon fibres. (Reproduced with permission from ref. 3. Copyright 1987 Gordon and Breach.)

increases from the untreated fibre to the coated fibres. As expected, the amphoteric coated fibres lead to the highest acid/base interaction with the amphoteric epoxy matrices.

Table IV. Specific Interaction Parameter A Between Carbon Fibre and Epoxy Matrix (in Arbitrary Units)

Matrix Fibre	Epoxy I	Epoxy II
Untreated	52	61
Oxidized 2	86	104
Oxidized 2'	88	-
Coated 3	152	188
Coated 4	127	-

Determination of Fibre-Matrix Adhesion. The average shear strength $\bar{\tau}$ and maximum shear strength τ_{max} at the fibre-matrix interface (7, 8, 29) are given by

$$\tau = \frac{d}{2 l_c} \sigma_f (l_c) \tag{17}$$

and

$$\tau_{max} = \frac{d\beta}{4} \frac{\sinh (\beta \frac{l_c}{2})}{\cosh (\beta \frac{l_c}{2}) - 1} \cdot \sigma_f(l_c) \tag{18}$$

with

$$\beta = \frac{2}{d} \left[\frac{E_m}{(1 + v_m) (E_f - E_m) \, Ln \, (\frac{2r_m}{d})} \right]^{1/2} \tag{19}$$

where d is the fibre diameter, r_m is the width of the single fibre-resin composite sample, E_m and E_f are the elastic moduli of the matrix and the fibre respectively, v_m is the Poisson's ratio of the matrix, and $\sigma_f (l_c)$ is the tensile strength of the fibre at the critical length l_c.

All these quantities are readily determined. However, the determination of l_c and $\sigma_f (l_c)$ necessitates a statistical analysis using the Weibull model. $\sigma_f (l_c)$ cannot be measured directly, since l_c is usually less than 0.5 mm. Therefore, it is determined from the tensile strength $\sigma_f (l)$ at higher gauge lengths using Equation 20

$$\sigma_f (l_c) = \sigma_f (l) \left(\frac{l}{l_c} \right)^{1/m} \tag{20}$$

where m is Weibull's shape parameter.

An example of results obtained with carbon fibres 1, 2, and 3 and resin I is shown in Table V.

Table V. Critical Length l_c of the Fibre in Matrix I and
Tensile Strengths of the Fibre

Fibre	l_c (mm)	σ_f (28 mm) (GPa)	σ_f (l_c) (GPa)
Untreated 1	0.50	3.02	5.9
Oxidized 2	0.41	3.19	5.9
Coated 3	0.36	2.89	6.3

The values of the average and the maximum shear strengths $\bar{\tau}$ and τ_{max} of the interface, as calculated from Equations 17 and 18, are presented in Table VI.

Table VI. Shear Strength of the Interface $\bar{\tau}$ and τ_{max};
Specific Interaction Parameter - Reversible Energy of Adhesion ($W_A^D + W_A^P$)

		$\bar{\tau}$ (MPa)	τ_{max} (MPa)	A (au)	W_A (mJ.m^{-2})
	Untreated 1	42	101	52	100
	Oxidized 2	50	113	86	103
Resin I	Oxidized 2'	47	110	88	-
	Coated 3	61	135	152	88
	Coated 4	54	128	127	-
	Untreated 1	18	37.5	61	109
Resin II	Oxidized 2	25	51.5	104	118
	Coated 3	32	65	188	111

It is observed that the fibre-matrix adhesion increases, whatever the nature of the epoxy, from the untreated to the oxidized and coated fibers. For instance, in the case of resin II, the increase is 70 to 80% when the coated fiber 3 is used.

Discussion

In previous work (30-32), it has been suggested that the adhesion between a carbon fibre and an epoxy matrix is essentially the result of physical bonds, either dispersive or polar. It is clear from the results in the last column of Table VI that there is no correlation between τ and the reversible energy of adhesion W_A, calculated as the sum of the dispersive and polar interactions at the fibre-matrix interface.

In contrast, as shown in Figures 8 and 9, there is a very good correlation between the interfacial shear strength τ (average or maximum) and the specific interaction parameter A, expressing the acid/base interactions exchanged at the interface.

The linear relationships obtained could mean that the interfacial adhesion results principally from acid/base or acceptor/donor interactions, assuming that the dispersive interactions are of the same magnitude. The meaning of the value of τ obtained by the intercept at the origin of the straight lines is not clear. It was first considered to be the contribution of dispersive interactions to the shear resistance of

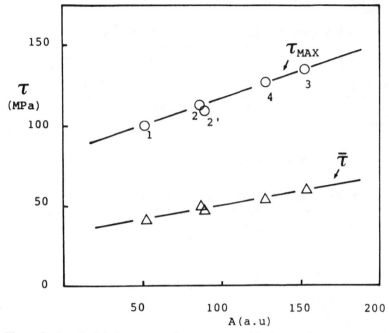

Figure 8: Interfacial shear strength τ versus specific interaction parameter A for the composites based on resin I.

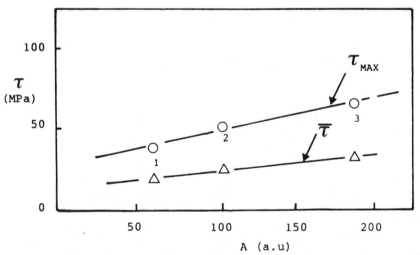

Figure 9: Interfacial shear strength τ versus specific interaction parameter A for the composites based on resin II.

the interface. However, it is now indicated that it is related to the mechanical properties of the epoxy matrix (33).

Although the correlation is quite convincing, acid/base interactions are not claimed to be the only explanation for the increased adhesion since many other mechanisms and phenomena, such as formation of an interphase, co-crosslinking, interdiffusion, mechanical anchoring and interfacial shrinkage could intervene .

Conclusion

Inverse gas chromatography at infinite dilution appears to be a powerful tool for studying the surface properties of carbon fibres and polymer matrices. The use of alkane probes and acid/base probes allows the characterization of the surfaces in terms of their London dispersive component of surface energy and their acid/base or acceptor/donor characteristics. A strong correlation was obtained between fibre-matrix adhesion, measured by a destructive fragmentation technique, and the level of acid base interactions calculated from the chromatographic analysis.

The concept of acid/base interactions constitutes an interesting, if not universal, approach to a better understanding of the interfacial properties of composite materials and could constitute a basis for a better choice of surface treatments applied to the fibres.

Literature Cited

1. Schultz, J. ; Lavielle, L. ; Simon, H. Proc. Intern. Symp. on Science and New Applications of Carbon Fibers, Toyohashi Univ., Japan, 1984, 3, 125.
2. Schultz, J. ; Lavielle, L. ; Martin, C. J. Chim. Phys, 1987, 84 (2), 23.
3. Schultz, J. ; Lavielle, L. ; Martin, C. J. Adhesion, 1987, 23, 45.
4. Schultz, J. ; Carré, A. ; Simon, H. Double Liaison, 1982, 322, 263.
5. Schultz, J. ; Tsutsumi, K. ; Donnet, J.B. J. Coll. Interf. Sci., 1977, 59 (2), 272 and 277.
6. Schultz, J. ; Cazeneuve, C. ; Shanahan, M. J. Adhesion, 1981, 12 (3), 221.
7. Kelly, A. ; Tyson, W.R. J. Mechanics and Physics of Solids, 1965, 13, 329.
8. Cox, H.L. Brit. J. Appl. Phys., 1952, 18, 273.
9. Weibull, W. Proc. Roy. Swedish Inst. Eng. Res., 1939, 151, 3.
10. Weibull, W. J. Appl. Mech., 1951, 18, 273.
11. Conder, J.R. ; Young, C.L. Physicochemical Measurements by Gas Chromatography, J. Wiley: New York, 1979.
12. Kemball, C .; Rideal, E.K. ; Proc. Roy. Soc., 1946, A 187, 53.
13. De Boer, J.H. ; Kruyer, S. ; Proc. K. Ned. Akad. Wet., 1952, B 55, 451.
14. Fowkes, F.M. Ind. Eng. Chem., 1964, 56 (12), 40.
15. Dorris, G.M. ; Gray, D.G. J. Coll. Interf. Sci., 1980, 77, 353.
16. Katz, S. ; Gray, D.G. J. Coll. Interf. Sci., 1981, 82, 318.
17. Anhang, J. ; Gray, D.G. J. Appl. Polym. Sci., 1982, 27, 71.
18. Drzal, L.T. ; Mescher, J.A. ; Hall, D.L. Carbon, 1979, 17, 375.
19. Li, S.K. ; Smith, R.P. ; Neumann, A.W. J. Adhesion, 1984, 17 (2), 105.
20. Vukov, A.J. ; Gray, D.G. Proc. 4th Intern. Carbon Conf. (Baden - Baden), 1986, 394.
21. Cazeneuve, C. ; Donnet, J.B. ; Schultz, J. ; Shanahan, M. Proc. 14th Biennial Carbon Conf., 1979, 216.
22. Drago, R.S. ; Vogel, G.C. ; Needham, T.E. J. Am. Chem. Soc., 1971, 93, 6014.
23. Drago, R.S. ; Parr, L.B. ; Chamberlain, C.S. J. Am. Chem. Soc., 1977, 99, 3203.

24. Gutmann, V. The Donor-Acceptor Approach to Molecular Interactions, Plenum Press, New York, 1983.
25. Fowkes, F.M. ; Maruchi, S. Org. Coatings Plast. Chem., 1977, 37, 605.
26. Fowkes, F.M. Rubber Chem. Tech., 1978, 57, 328.
27. Fowkes, F.M. J. Adhesion Sci. Tech., 1987, 1, 7.
28. Saint-Flour, C. ; Papirer, E. Ind. Eng. Chem. Prod. Res. Dev., 1982, 21, 666.
29. Fraser, W.A. ; Ancker, F.H. ; Di Benedetto, A.T. ; Elbirli, E. Polym. Composites, 1983, 4 (4), 234.
30. Fitzer, E. ; Geigl, K.H. ; Huettner, W. ; Weiss, R. Carbon, 1980, 18, 389.
31. Drzal, L.T. ; Rich, M.J. ; Lloyd, P.F. J. Adhesion, 1982, 16, 1.
32. Morita, K. ; Murata, Y. ; Ishitani, A. ; Murayama, K. ; Ono, T. ; Nakajima, A. Pure Appl. Chem., 1986, 58 (3), 455.
33. Martin, C. Ph. D. Thesis, Université de Haute-Alsace, Mulhouse, France, 1988.

RECEIVED November 2, 1988

Chapter 15

Surface Energetics of Plasma-Treated Carbon Fiber

Sheldon P. Wesson[1] and Ronald E. Allred[2]

[1]TRI/Princeton, P.O. Box 625, Princeton, NJ 08542
[2]PDA Engineering, Materials Development Department, 3754 Hawkins NE, Albuquerque, NM 87109

Carbon fiber surfaces were analyzed by inverse gas chromatography, using Lewis acids and bases as probe adsorbates to determine the effect of radio frequency glow discharge plasma treatments on carbon surface energetics. Adsorption isotherms computed from chromatograms were analyzed with the CAEDMON algorithm (Computed Adsorption Energy Distribution in the MONolayer) to obtain histograms of surface area fraction versus adsorption energy. Changes in surface heterogeneity revealed by adsorptive energy distributions were correlated with wetting data, and surface chemical composition deduced from high resolution x–ray photoelectron spectroscopy.

Impetus to study the surface energetics of carbon fiber arises from efforts to improve physical properties of high performance composites by manipulating the nature and extent of interfacial interaction between the reinforcement and the matrix polymer. Examination of changes in surface energetics that result when carbon fiber is subjected to chemical or thermal treatment provides a useful intermediate response between reinforcement processing and composite testing: surface energetics analysis can be used to screen treatment recipes for a variety of reinforcement/matrix combinations.

A more vital application is to discern how reinforcement surface treatments improve adhesion to thermoplastic matrices. Since the nonreactive nature of thermoplastics normally precludes interfacial covalent bond formation, secondary bonding forces, such as London dispersion interactions and Lewis acid/base interactions, may play a major role in these circumstances. These secondary binding forces are subject to surface energetics analysis.

Most surfaces are not energetically uniform; they feature sites with differing modes of interaction with other molecules. Surface heterogeneity, in the present context, indicates a range in the nature and magnitude of attraction at these sites. This attribute can be studied via measurements of solid/gas or solid/liquid interaction. Surface heterogeneity can be assessed from wetting hysteresis, for example, but the analysis is not quantitative except under carefully contrived circumstances. A liquid meniscus wetting a fiber perimeter interacts simultaneously with all of the sites on the three–phase boundary; therefore, one measurement of adhesion tension obtained at any position is the average of many values.

0097–6156/89/0391–0203$06.00/0
© 1989 American Chemical Society

Gas molecules are not so constrained, interacting preferentially with portions of the surface that present relatively strong attractive forces. Adsorbate molecules cover these sites at lower equilibrium pressures than those required for sites that present weaker attractions. The adsorption isotherm (moles adsorbed per gram of solid versus adsorbate pressure) can be analyzed to obtain a histogram of surface area fraction versus adsorptive energy.

The objectives of this paper are to demonstrate how monolayer adsorption isotherms can be obtained on carbon fiber surfaces by inverse gas chromatography (IGC), and to compare results of solid/gas adsorption with those of solid/liquid wetting. This information is correlated with independent assessments of surface chemical functionality provided by wet chemical titrations and x–ray photoelectron spectroscopy (XPS).

Three carbon fiber surfaces are compared in this study: Amoco Thornel–300 graphitized at 2500° C, Hercules IM6 carbon fiber subjected to surface oxidation but with no subsequent deposition of size (oils, surfactants, or polymer films applied to facilitate manufacture and processing), and unsized IM6 treated with a radio frequency glow discharge plasma.

Background

Whereas conventional chromatographic methods manipulate the surface energetics of sorbents to separate fluid mixtures, inverse gas chromatography uses known properties of fluids to characterize surface properties of solids. Specifically, Lewis acids and bases are used, in this study, as probes to deduce the nature and extent of solid/gas attraction from the shape of chromatograms, which are transformed adsorption isotherms. IGC can determine the specific surface (m^2/g) of the substrate, whether the surface is acidic, basic, amphoteric, or neutral, and whether the surface is homogeneous or heterogeneous.

This analysis is based on the relationship between the retention time of the probe vapor in the column and the Henry's law constant (vapor partition coefficient) that relates mole fraction adsorbed with equilibrium pressure ([1]). The connection between retention time, a term that evokes kinetic properties, and an equilibrium constant is not necessarily obvious, but the following analysis shows the retention time to be the Henry's law constant K multiplied by calibration factors ([2]). At low equilibrium pressures,

$$K = \frac{n}{p} \qquad (1)$$

where n is the moles adsorbed per gram of solid g, and p is the adsorbate gas pressure. If n_{ads} is the total moles adsorbed, n_{gas} is the moles in the gas phase, and V_{col} is the dead volume in the chromatographic column,

$$K = \frac{n_{ads}/g}{n_{gas}RT/V_{col}} = \frac{n_{ads}}{n_{gas}} \frac{V_{col}}{gRT} \qquad (2)$$

so that K is proportional to the ratio of moles adsorbed to moles in the vapor phase at equilibrium. Let t_r be the time for the adsorbate to pass through the column, and let t_m be the time of passage for an inert molecule. Then t_r is t_m divided by the fraction of solute molecules in the gas phase: a molecule that spends half its time in the mobile phase takes twice as long to traverse the column as one that spends all of its time there.

$$t_r = \frac{t_m}{\left[\dfrac{n_{gas}}{n_{gas} + n_{ads}}\right]} = t_m \left[1 + \frac{n_{ads}}{n_{gas}}\right] \tag{3}$$

and

$$\frac{n_{ads}}{n_{gas}} = \frac{t_r - t_m}{t_m} = \frac{t_a}{t_m} \tag{4}$$

The adjusted retention time t_a is proportional to the Henry's law constant, as shown by substituting Equation 4 into Equation 2:

$$K = \frac{V_{col}}{gRT} \frac{t_a}{t_m} = \frac{F}{gRT} t_a \tag{5}$$

where F, the carrier gas flow rate, is substituted for V_{col}/t_m .

Isotherms from Chromatograms. Consider the process of injecting a column with n_{inj} moles of vapor that are reversibly adsorbed. The determination of adsorbate pressure from the chromatogram requires a linear detector response v to f, the rate at which adsorbate enters the detector:

$$v = xf \tag{6}$$

In the absence of irreversible adsorption, the moles of adsorbate that enter the detector during time t are given by

$$n_{inj} = \int f \, dt \tag{7}$$

and the calibration factor x is found by measuring the peak area A :

$$A = \int v \, dt = \int xf \, dt = x \, n_{inj} \tag{8}$$

The concentration c of adsorbate in the carrier gas is expressed in terms of the detector response and the carrier flow rate by substituting Equation 8 into Equation 6:

$$c = \frac{f}{F} = \frac{n_{inj}}{AF} v \tag{9}$$

If ideal gas behavior is assumed, the adsorbate pressure is given by

$$p = \frac{n_{inj} RT}{AF} v \tag{10}$$

The moles adsorbed per gram of solid is obtained from Equation 1 by substituting Equation 5 for K and Equation 10 for p :

$$n = Kp = \frac{n_{inj}}{gA} t_a v \qquad (11)$$

This analysis is used with a method for determining an isotherm from one injection of adsorbate, termed "Elution by Characteristic Point" (3). The method requires that chromatograms from a series of injections of varying adsorbate volume be in registry along the diffuse profile (the portion of the chromatogram after peak maximum), and that there be no inflection point in the relevant portion of the isotherm (4). If the diffuse profiles are congruent and the peak fronts are sharp (the chromatograms begin abruptly), showing that maximum coverage is below any inflection point, the isotherm can be determined by computing n and p at many points along the diffuse profile of the largest peak (5).

Analysis of Heterogeneity. The monolayer analysis consists of three elements: an adsorption isotherm equation, a model for heterogeneous surfaces, and an algorithm such as CAEDMON, which uses the first two elements to extract the adsorptive energy distribution and the specific surface from isotherm data. Morrison and Ross developed a virial isotherm equation for a mobile film of adsorbed gas at submonolayer coverage (6):

$$\ln p = \ln \frac{n}{\Sigma} + \frac{2nB^*}{\Sigma} + \frac{3n^2 C^*}{2\Sigma^2} + \ldots + \ln K \qquad (12)$$

where K is an integration constant, and B^* and C^* are reduced virial coefficients that describe adsorbate–adsorbate interaction in two dimensions. The specific surface Σ is given by

$$\Sigma = n_m N_a \sigma^2 \qquad (13)$$

where n_m is moles adsorbed per gram at full monolayer coverage, N_a is Avogadro's number, and σ is the Lennard–Jones distance of closest approach for the adsorbate molecule. The integration constant in Equation 12 is equivalent to Σ/K, and can be expressed in terms of U_0, the adsorptive potential of the solid surface:

$$K = A^0 e^{-U_0/RT} \qquad (14)$$

where A^0 describes the difference in degrees of freedom between adsorbate molecules in the bulk gas and the adsorbed film. A^0 is not evaluated in the present analysis. Ross and Morrison treat the heterogeneous surface as a collection of monoenergetic patches, with different adsorptive energies (7). Patches are filled simultaneously, though not to the same density, subject to the condition that the adsorbed phase on each patch has the same chemical potential. An arbitrary range of relative adsorption energies $-RT \ln K_i$ (kJ/mole) is selected in a systematic manner (8), where K_i is the Henry's law constant for a given patch. The number of moles adsorbed per gram at each pressure is obtained by summing the individual values of the number of moles per gram adsorbed on each patch n_i, over all the patches (9).

$$n(p) = \sum_i n_i(p) = N_a \sigma^2 \sum_i n_{mi}\, f(p/K_i) \qquad (15)$$

where n_{mi} is the moles adsorbed per gram at full monolayer coverage on a given patch, and the function $f(p/K_i)$ is found by solving Equation 12 iteratively for n_i/Σ_i. A linear operations algorithm provides a set of non–negative n_{mi} that minimizes the relative deviation between the model isotherm and experimental data ($\underline{10}$). The sum of the specific surface areas of all the patches is the specific surface of the solid, and the adsorptive energy histogram is obtained as a plot of Σ_i versus $-RT \ln K_i$.

Materials and Methods.

XPS. X–ray photoelectron spectra were obtained by Rocky Mountain Laboratories, Inc., Golden, CO, using a Surface Sciences SSX–100 spectrometer with an Al $K\alpha$ source. High resolution C_{1s} peaks were averaged over 10 to 15 scans using a spot size of 300 μm^2 with no flood gun. O_{1s} peaks were averaged over 30 scans. A Gaussian curve fitting routine was used to resolve high resolution photopeaks into components based on binding energy references from model compounds ($\underline{11}$).

Titrations. Carbon fiber specimens were dried under vacuum (5 kPa) for one hour at 50° C prior to titration in benzene with 1,2–diphenylguanidine for acid groups, and diphenyl phosphate for basic groups ($\underline{12}$). Fibers were reacted for 15 to 30 minutes with the appropriate acid or base solution prior to back titration. The indicator range of bromophthalein Magenta E (pH 3.0 to 4.2) was corrected with blank titrations.

Wetting. Fiber wettability was measured by the Wilhelmy technique ($\underline{13}$) using methylene iodide (nonpolar oil), formamide (Lewis base), and ethylene glycol (Lewis acid). Carbon monofilaments were glued to metal hooks and suspended vertically from the working arm of a Cahn 2000 microbalance, while a precision elevator raised and lowered a liquid surface along 15 mm of fiber. A computer periodically recorded the change in apparent mass caused by wetting forces at the three–phase boundary. Advancing work of adhesion is computed for each value of apparent mass W as

$$W = \gamma_1 + \frac{Wg}{\pi d} \qquad (16)$$

where γ_1 is the liquid surface tension, d is the fiber diameter computed from the known adhesion tension in a low energy liquid, and g is the acceleration due to gravity. Surface tension was measured at 50.1 mN/m for methylene iodide, and is taken as 58.2 mN/m for formamide ($\underline{14}$), and 47.7 mN/m for ethylene glycol ($\underline{15}$). The dispersion force component of the solid surface energy is computed from the advancing wettability in methylene iodide W_{oil} as

$$\gamma_s^d = \frac{W_{oil}^2}{4\gamma_{oil}} \qquad (17)$$

The dispersion force component of the advancing work of adhesion is found as

$$W^d = 2(\gamma_s^d \, \gamma_l^d)^{1/2} \tag{18}$$

where the dispersion force component of liquid surface tension is 39.5 mN/m for formamide ($\underline{14}$), and 30.1 mN/m for ethylene glycol ($\underline{15}$). The acid/base component of the work of adhesion is given by

$$W^{a/b} = W - W^d \tag{19}$$

Inverse Gas Chromatography. Columns were constructed from 1/4–inch stainless steel tubing (5.2 mm inner diameter) with passivated inner walls supplied by Supelco Corp. Sections 55 cm long were fitted with Swagelok nuts and ferrules, and then weighed. Fiber yarns were wound into 60 cm loops that were drawn into the tubes and cut flush at the ends; the columns were then weighed again. Sample weights varied from 2 to 3 g.

A Hewlett–Packard 5880A gas chromatograph fitted with a flame ionization detector and a D/A output board was connected to a 20 MHz 80386/80387 Micronics computer containing a 16 bit Data Translation Series 2801 A/D conversion card. Routines written in ASYST programming language collected detector voltages at frequencies between 1 and 10 Hz.

Injector and detector temperatures were maintained at 150°C. Nitrogen carrier flow rates were measured with a Gasmet flow meter and were maintained between 22 and 24 ml/min. Measured flow rates **F** were corrected for column temperature using

$$F = \mathbf{F}\,\frac{T_{col}}{T_{amb}} \tag{20}$$

There was no pressure drop across these columns, obviating the need for the James/Martin correction ($\underline{16}$). Gas holdup times were measured with 10 μl injections of methane. Hamilton 7101NCH syringes were used to inject 2 μl volumes of pentane (neutral probe) and t–butylamine (Lewis base); injections of t–butanol (Lewis acid) were restricted to 0.5 μl in order to obtain sharp peak fronts. Three chromatograms were obtained on each column with each probe at 30°C after conditioning the column overnight at 50°C.

Chromatograms were obtained at the minimum attenuation a required to maintain the detector signal below saturation at peak maximum; zero attenuation was implemented after the signal fell below $1/a$ of saturation, permitting desorption at low coverage to be measured with maximum resolution. Chromatograms using pentane, t–butylamine and t–butanol were collected over periods of 1000, 6000, and 2000 seconds, respectively. An additional two hours elapsed between measurements with t–butylamine to allow chemisorbed probe to clear from the column.

Results and Discussion.

XPS. Surface concentrations of carbon, oxygen, and nitrogen for the three fiber specimens are listed in Table I. Results are normalized to $C + O + N = 100$ atom percent. Trace concentrations of Si and Al present on the IM6 samples are not tabulated.

Table I. Surface Composition of Carbon and Graphite Fibers

Substrate	at% C	at% O	at% N
Graphitized Thornel–300	98.4	1.6	nd[†]
Unsized IM6	89.9	7.9	2.2
Unsized IM6, acid plasma	83.4	13.8	2.8

[†] not detected

These survey spectra show that the graphitized Thornel–300 surface is composed of carbon and a small amount of oxygen. IM6 fiber surfaces contain larger quantities of oxygen and small amounts of nitrogen (probably residual species from the polyacrylonitrile fiber precursor). IM6 fiber surfaces acquire a significant oxygen concentration during manufacture. Acid plasma treatment increases the surface oxygen concentration by augmenting the oxygen species present at crystallite edges and defects (17, 18), and possibly by attacking the graphite basal plane. Components of high resolution C_{1s} and O_{1s} envelopes are presented in Table II.

Table II. Components of C_{1s} and O_{1s} Photopeaks

Binding Energy, eV	Species	Graphitized Thornel–300	Unsized IM6	Unsized IM6, acid plasma
C_{1s}:		percent of C_{1s} photopeak		
284.6	primary carbon	67.8	59.9	56.5
285.3	β–carbon in conjugated systems	14.4	16.6	18.4
286.4	hydroxyl, ether, ester single bond	7.1	10.9	9.4
287.6	ketone, amide	3.6	2.7	5.2
288.7	carboxyl, ester double bond	nd	2.6	6.5
290.7	aromatic shakeup, carbonate	7.1	7.3	4.0
O_{1s}:		percent of O_{1s} photopeak		
531.6	ketone, amide	55.5	52.0	38.8
532.8	carboxyl, ester double bond; hydroxyl, ether single bond			45.5
		44.5 } nr[‡]	48.0	
533.8	carboxyl, ester single bond			15.7

[‡] not resolved

Analysis of C_{1s} photopeak components at 286.4 eV and 287.6 eV indicates that the small amount of oxygen on graphitized Thornel–300 is present primarily as hydroxyl, ether, and ketone groups. A strong shakeup peak from the highly graphitic structure is present at 290.7 eV. Unsized IM6 fiber is less graphitic than the Thornel–300 specimen and should show a smaller shakeup peak; the substantial peak at 290.7 eV is partially attributed to carbonate functionality, which also contributes to the peak at 286.4 eV. The appearance of a peak component at 288.7 eV on unsized IM6 shows that oxidation during fiber manufacture implants some carboxyl functionality (19). The component at 288.7 eV is substantially larger on plasma treated surfaces; this observation, together with the appearance of an O_{1s} peak component at 533.8 eV, shows that additional oxygen implanted by plasma treatment is present primarily as carboxyl groups. The peak at 290.7 eV is substantially diminished, but the component at 286.4 eV is unchanged, suggesting that acid plasma converts carbonate structures to hydroxyl as well as carboxyl functionality. The main inference is that graphitized Thornel–300, unsized IM6, and plasma treated IM6 substrates constitute a graded series in increasing surface acidity.

Titrations. Values for surface site concentration listed in Table III are averages of titrations performed in triplicate.

Table III. Surface Site Titrations

Substrate	Acid Sites/nm²	Basic Sites/nm²
Graphitized Thornel–300	0.4	0.8
Unsized IM6	1.3	0.5
Unsized IM6, acid plasma	2.1	nd

Graphitized Thornel–300 fiber shows minimal surface acid concentration. Plasma treatment increases the acidity of unsized IM6 fiber substantially, and diminishes its basicity below detectable levels. These results corroborate the central inference from XPS analysis, i.e., that the oxygen added by plasma treatment is present primarily as hydroxyl and carboxyl functionalities.

Wetting. Advancing work of adhesion for methylene iodide, formamide, and ethylene glycol on the three carbon fiber specimens is presented in Table IV.

Table IV. Carbon Fiber Wettability, mN/m

	CH_2I_2		$HCONH_2$		$HOCH_2CH_2OH$	
Substrate	W	γ_s^d	W	$W^{a/b}$	W	$W^{a/b}$
Graphitized Thornel–300	88.6	39.1	80.8	2.2	75.6	7.0
Unsized IM6	84.4	35.5	90.6	15.7	83.7	18.3
Unsized IM6, acid plasma	86.7	37.4	113.8	36.8	85.3	18.2

Values of V are the average of 180 measurements per fiber using three to seven fibers for each solid/liquid combination; the maximum 95% confidence interval is 2 mN/m. V for wetting with methylene iodide is equivalent for the three fiber specimens. Values of γ^d are in accord with literature values for similar substrates from wetting measurements (20–23), and are less than some determinations obtained by IGC (24, 25). All of these values are significantly lower than 113 mN/m obtained for heptane on clean graphite (26). Low values of γ^d from wetting carbon fibers may be caused by polar groups on the surface, organic surface contaminants, surface rugosity, or π_e, the mitigation of surface energy by adsorption of ambient vapor (27, 28).

$V^{a/b}$ for wetting with formamide (the Lewis base) is negligible for graphitized Thornel–300. Unsized IM6 shows significant acid/base interaction with formamide; the acid plasma treated surface shows the highest level of acid/base attraction with the basic probe. This trend is in accord with the XPS analysis and titration results. Graphitized Thornel–300 shows a low but significant level of acid/base interaction with ethylene glycol (the acid probe). IM6 fibers show significantly greater levels of $V^{a/b}$ for wetting with the Lewis acid, an observation that is not predicted by XPS or titration data. (Ethylene glycol and formamide spread on many of the IM6 fibers in the receding mode, obviating an evaluation of wetting hysteresis.)

Inverse Gas Chromatography. Diffuse profiles of chromatograms were transformed to isotherms by using Equation 11 to find n for each v and t_a, and Equation 10 to find p for each v, at every point on the chromatogram between peak maximum and the end of the tail. Points near peak maximum were discarded as being influenced by diffusion and other extraneous effects. Isotherms in Figure 1 are displayed in log/log format to emphasize differences at low pressure and coverage. The geometric specific surface $\Sigma_{geo} = 4/\rho d$ was computed for each carbon fiber sample using the density ρ supplied by the manufacturer, and the diameter found from wetting measurements, as listed in Table V. Adsorption volumes are expressed as n/Σ_{geo} ($\mu moles/m^2$) to normalize the isotherms for all experimental conditions except differences in adsorption energetics. 10 $\mu moles/m^2$ is approximately one monolayer for these adsorbates. Equilibrium pressure is given in kiloPascals.

Table V. Properties of Graphite and Carbon Fiber

Substrate	ρ (g/cm^3)	d (μm)	Σ_{geo} (m^2/g)
Graphitized Thornel–300	1.76	6.8	0.33
Unsized IM6	1.73	5.3	0.44

The shapes of these isotherms are controlled by surface heterogeneity, and thus provide a means of deducing the various modes of interfacial attraction presented by the substrates. Isotherms of pentane on unsized IM6 before and after plasma treatment are congruent. Pentane interacts only by dispersion force attraction, as does methylene iodide; the registry of these isotherms is predicted by the similarity in γ^d obtained from wetting measurements. Pentane isotherms serve as a reference with which to assess the effect of additional modes of solid/vapor interaction with acidic and basic probes.

Isotherms for adsorption of t–butanol (the Lewis acid) on both IM6 substrates are in registry within experimental precision.   The adsorptive capacity of carbon fiber for the Lewis acid is greater than for pentane across the entire pressure range, indicating that dispersion force interaction is augmented with weak attraction to basic sites on the substrate.   These observations corroborate trends evinced by wetting with ethylene glycol.   Isotherms for t–butylamine adsorption are in registry above 10^{-4} kPa; the plasma treated surface has a greater adsorptive capacity for the basic probe at lower pressures. The adsorptive capacity of both specimens for the Lewis base is more than an order of magnitude greater than for acidic or neutral probes at pressures below 10^{-4} kPa.   The main inferences are that both IM6 surfaces are acidic, and plasma treatment increases the acid site concentration imparted during manufacture, which is in accord with the preceding analysis of XPS spectra, titrations, and wetting measurements.

Observations about surface energetics can be quantified by applying the CAEDMON algorithm to adsorption data.   Isotherms for t–butylamine on graphitized Thornel–300, unsized IM6, and unsized IM6 subjected to plasma treatment were analyzed to obtain adsorption energy profiles.   Parameters used in the CAEDMON analysis are presented in Table VI.

Table VI. Parameters for Gas Adsorption Analysis

Adsorbate	ϵ/k (°K)	σ (nm)
pentane	345 (29)	0.5769 (29)
t–butylamine	310 (30)	0.5701 (31)
t–butanol	450 est	0.5950 (31)

The histograms displayed in Figure 2 are generated as follows: compute the reduced temperature $T* = kT/\epsilon$ using the Lennard–Jones force constant for t–butylamine listed in Table VI; use $T*$ to interpolate for reduced virial coefficients $B*$ and $C*$ from values calculated by Morrison and Ross (6); submit the virial coefficients, a vector of adsorptive energies, and a vector of specific moles adsorbed from the isotherm to the CAEDMON algorithm; then compute Σ_i for each n_{mi} in the vector returned by CAEDMON, using σ from Table VI in Equation 13. The histogram from adsorption of t–butylamine on graphitized Thornel–300 shows a narrow range of low energy sites; more than half of the surface comprises one patch (small boxes on the abcissa are energy values submitted to the CAEDMON algorithm).

Oxidation during manufacture creates heterogeneity, manifested as a broader range of low to medium energy sites on the unsized IM6 surface. These weak acid sites increase the adsorptive capacity of IM6 for the basic probe at pressures above 10^{-4} kPa, compared with the reference for dispersion force interaction denoted by pentane isotherms (see Figure 1). The most important phenomenon is the appearance of a patch with a very high adsorptive energy, constituting less than 5% of the specific surface. Plasma treatment does not change the distribution of weak acid sites appreciably (t–butylamine isotherms are congruent above 10^{-4} kPa), but increases the surface concentration of high energy sites to about 7% of the specific surface. These high energy patches are similar in magnitude and proportion to C_{1s} peak components at 288.7 eV.

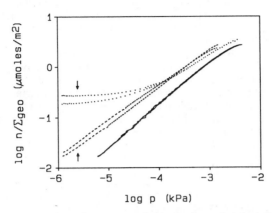

Figure 1. Pentane (lines), *t*–butanol (dashes), and *t*–butylamine (dots), on unsized IM6 fiber at 30° C. Arrows denote the acid plasma treated surface.

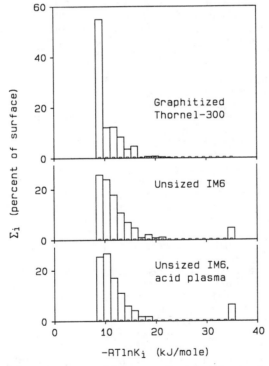

Figure 2. Adsorptive energy histograms for *t*–butylamine isotherms at 30° C.

The range of adsorptive energies supplied to the algorithm affects the shape of the histogram and the computed value of specific surface. Σ is sensitive to the position of the lowest energy patch; the low energy bound is adjusted to bring Σ to within 10% of Σ_{geo} in the present analysis. (An empirical procedure is used when Σ_{geo} is not known (8), a mechanism employed in lieu of viable theory for computing the monolayer/multilayer transition on heterogeneous substrates.) The exact location of the highest site energy is unimportant when the low pressure sector of the isotherm corresponds to the Henry's law region of coverage on the highest energy patch. This is the case for adsorption on graphitized Thornel–300, as shown by the assignment of zero specific surface to the highest site energies.

The last increment of t–butylamine takes several hours to traverse the column when IM6 surfaces are eluted. This is evidence of very strong solid/vapor interaction: in these circumstances the highest site energy is indeterminate in principle, because this patch saturates at pressures much lower than the operating range of the detector. The position of the highest energy patch is arbitrarily assigned as the lowest energy that brings the computed fit into registry with the low pressure sector of the isotherm. The patch area thus obtained is an upper limit, and a good approximation because it will not vary much if its energy assignment is modified slightly. (The fit will fall out of registry with the data if the energy assignment is too high.)

The analysis of chemisorptive patches lies within the scope of the CAEDMON procedure, since a model for localized adsorption such as the Fowler–Guggenheim equation (32) can be used to compute the high energy sector of the distribution. Isotherms for the Lewis base on IM6 surfaces show zero slope at pressures below 10^{-5} kPa, however, indicating that the adsorbate complexes with chemisorptive sites on the substrate. No benefit is achieved by refining the model unless the stoichiometry of complex formation is determined. This can be assessed by programmed thermal desorption, which can also provide an independent estimate of effective bond strength if multiple measurements are obtained at varying rates of heating (33).

The present analysis attributes skew in chromatograms only to heterogeneity. Adsorptive energy distributions presented above are distorted by processes that skew chromatograms but are unrelated to surface energetics. Residual acid sites on the inner walls of steel columns will distort diffuse profiles of chromatograms obtained with basic probes. This effect is small when the tubing is passivated by organosilane deposition, but it can be eliminated entirely by using nickel tubing for column construction.

Diffusion and related phenomena that cause peak broadening have their greatest impact on chromatograms near peak maximum, and therefore on isotherms for relatively homogeneous substrates. The assumption that effects of diffusion at the front and back of the peak are equivalent permits subtracting each point on the peak front from the corresponding detector response on the diffuse branch (34). Skewed chromatograms often show sharp peak fronts, however; in these circumstances, diffusion effects can be corrected only by modeling the chromatographic process (35–37).

Boudreau and Cooper showed that adsorptive energy distributions can be computed directly from chromatograms (38). Their method is similar in principle to the present analysis, except for the use of an adsorption isotherm equation that provides an analytical solution for Equation 15. This simplification, which describes all adsorption isotherms as if they occur below the two–dimensional critical temperature $T^*_c = 0.60$ (6), speeds computation of the histogram. Its practical effect is to broaden the energy scale artificially, however, particularly when light vapors are used to characterize low energy, homogeneous substrates.

Conclusions. Results of XPS analysis, surface titration, wetting, and CAEDMON analysis of adsorption isotherms are summarized in Table VII.

Table VII. Surface Properties of Graphite and Carbon Fiber

Measurement	Graphitized Thornel–300	Unsized IM6	Unsized IM6, acid plasma
surface O (at%)	1.6	7.9	13.8
C_{1s}: 288.7 eV	nd	2.6	6.5
acid sites/nm²	0.4	1.3	2.1
$\gamma^{a/b}$ HCONH₂ (mN/m)	2.2	15.7	36.8
$\Sigma_{chemisorption}$ (%)	nd	4.7	6.4

Trends in surface composition, functionality, and energetics shown in Table VII are in good qualitative accord. It is apparent, however, that these techniques are sensitive to different aspects of surface acidity. A stoichiometry of 1:1 for chemisorptive complex formation with 2.1 acid sites/nm² on plasma treated IM6 corresponds to a coverage of 60% with t-butylamine, an order of magnitude greater than $\Sigma_{chemisorption} = 6.4\%$ from IGC measurements. This suggests that titration is sensitive to a range of weak acid functionality in addition to high energy sites, as is wetting with formamide, since both measurements vary almost quantitatively with surface oxygen concentration.

The most important inference is that $\Sigma_{chemisorption}$ is a direct response to carboxyl group concentration indicated by the XPS photopeak component at 288.7 eV. It seems likely that weak acid functionality is of minor import to applications for surface treatments, while interfacial phenomena such as practical adhesion may be sensitive to small concentrations of very high site energies. Interphase modification in epoxy resins, for example, can occur by direct reaction of epoxide groups with surface carboxyls (17), or by accelerated cure chemistry near the surface (39). Carboxyl groups on carbon surfaces may interact with basic moieties in polymers such as polycarbonate or poly(ethylene)oxide (40–42), or promote interfacial crystallinity that improves impact strength and other aspects of composite performance (43, 44).

Acknowledgments.

This work was supported in part by NASA–LARC under contract NAS1–18469, monitored by Dr. Jeffrey Hinkley.

Literature Cited.

1.   Ross, S; Olivier, J. P. On Physical Adsorption; Wiley–Interscience: New York, 1964; p 8.
2.   Meyer, M. F. J. Chem. Ed. 1980, 57 (2), 120.
3.   Huber, J. F. K; Keulemans, A. I. M. In Gas Chromatography 1962; M. van Swaay, Ed.; Butterworth: London, 1962; p 24.

4.   Conder, J. R.; Young, C. L. Physicochemical Measurement by Gas
     Chromatography; Wiley–Interscience: New York, 1979; p. 44.
5.   Neumann, M. G. J. Chem. Ed. 1976, 53 (11), 708.
6.   Morrison, I. D.; Ross, S. Surface Sci. 1973, 39 (1), 21.
7.   Ross, S.; Morrison, I. D. Surface Sci. 1975, 52 (1), 103.
8.   Wesson, S. P.; Vajo, J. J.; Ross, S. J. Colloid Interface Sci.
     1983, 94 (2), 552.
9.   Sacher, R. S.; Morrison, I. D. J. Colloid Interface Sci. 1979, 70 (1), 153.
10. Hanson, R. J. Solving Least Squares Problems; Prentice Hall:
     Englewood Cliffs, New Jersey, 1974.
11. Clark, D. T. In Characterization of Metal and Polymer Surfaces;
     L–H Lee Ed.; Academic Press: New York, 1977; Vol. 2, p 5.
12. Wilson, K. V.; Harrah, L. A. PDA Engineering: Unpublished data.
13. Wesson, S. P.; Jen, J. S. Proc. 16th National SAMPE Tech. Conf.
     1984, 16, 375.
14. Fowkes, F. M. In Treatise on Adhesion and Adhesives; Patrick, R. L.
     Ed.; Marcel Dekker: New York, 1967; p 360.
15. Wakida, T.; Kawamura, H.; Song, J. Sen–I Gakkaishi 1987, 32 (7), 384.
16. James, A. T.; Martin, A. J. P. J. Biochem. 1952, 50, 679.
17. Kozlowski, C.; Sherwood, P. M. A. Carbon 1987, 25 (6), 751.
18. Galuska, A. A.; Madden, H.; Allred, R. Appl. Surface Sci. 1988, 32, 253.
19. Donnet, J. B.; Ehruburger Carbon 1977, 15, 143.
20. Donnet, J. B.; Brendle, M. Carbon 1986, 24 (6), 757.
21. Kaelble, D. H.; Dynes, P. J.; Maus, L. J. Adhesion 1974, 6, 239.
22. Drzal, L. T.; Mescher, J. A.; Hall, D. L. Carbon 1979, 17 (5/A), 375.
23. Hammer, G. E.; Drzal, L. T. Appl. Surface Sci. 1980, 4, 340.
24. Shultz, J.; Lavielle, L.; Martin, C. J. Adhesion 1987, 23, 45.
25. Vukov, A. J.; Gray, D. G. Langmuir 1988, 4 (3), 743.
26. Fowkes, F. M. In Chemistry and Physics of Interfaces; Gushee, D. E.
     Ed.; ACS Publications: Washington, D. C., 1975; p 11.
27. Fowkes, F. M. J. Colloid Interface Sci. 1980, 78, 200.
28. Ross, S.; Morrison, I. D. Colloidal Systems and Interfaces;
     Wiley–Interscience: New York, 1988; p 92.
29. Hirchfelder, J. O.; Curtiss, C. F.; Bird, R. B. Molecular Theory of Gases
     and Liquids; Wiley: New York, 1964; p 1112.
30. DIPPR Data Compendium of Pure Compound Properties: NBS Standard
     Reference Database No. 11.
31. McClellan, A. L.; Harnsberger, H. J. Colloid Interface Sci. 1967, 23, 557.
32. Fowler, R. H.; Guggenheim, E. A. Statistical Thermodynamics;
     Cambridge Press: London, 1949; p 442.
33. Nishioka, G.; Schramke, J. A. J. Colloid Interface Sci. 1985, 105 (1), 102.
34. Dollimore, D.; Heal, G. R.; Martin, D. R. J. Chromatogr. 1970, 50, 209.
35. Pawlisch, C. A.; Macris, A.; Lawrence, R. L. Macromolecules
     1987, 20 (7), 1564.
36. Pawlisch, C. A.; Brick, J. R.; Lawrence, R. L. Macromolecules
     1988, 21 (6), 1685.
37. Munk, P.; Hattam, P. Macromolecules 1988, 21 (7), 2083.
38. Boudreau, S. P.; Cooper, W. T. Anal. Chem. 1987, 59, 353.
39. Garton, A.; Stevenson, W. T. K.; Wang, S. P. J. Polym. Sci. Polym.
     Chem. Ed. 1988, 26, 1377.
40. Manson, J. A. Pure & Appl. Chem. 1985, 57 (11), 1667.
41. Valia, D. A. Doctoral dissertation, Lehigh University, 1987.
42. Fowkes, F. M. J. Adhesion Sci. Tech. 1987, 1 (1), 7.
43. Hartness, J. T. SAMPE J. 1984, 20 (5), 6.
44. Lee, Y.; Porter, R. S. Polym. Eng. Sci. 1986, 26 (9), 633.

RECEIVED November 2, 1988

Chapter 16

Determination of Fiber–Matrix Adhesion and Acid–Base Interactions

A. E. Bolvari and Thomas Carl Ward[1]

Department of Chemistry and Adhesion Science Center, Virginia Polytechnic Institute and State University, Blacksburg, VA 24061

Inverse gas chromatography (IGC) was used to determine the dispersive and non-dispersive (acid/base) surface energies of carbon fibers. Further investigations were carried out on thermoplastic polymers using capillary column IGC. Substantiating information concerning the chemical composition of the fibers was obtained using X-ray photoelectron spectroscopy (XPS). A single fiber critical length test was used to correlate the interfacial adhesion of the fiber and polymer matrix to the nature of their surfaces.

Thermoplastic polymers that have the necessary requirements to qualify as a matrix in composite structural components (for example, solvent resistance, high modulus, high glass transition, and good fracture energy) tend to exhibit poor adhesion to carbon fibers. The weakness of this fiber–matrix interface results in a composite that may be unacceptable in its final performance. The question of why this bond is weak and concern for quantifying the relationship with respect to adhesion led to the current investigation.

There are four general models for of adhesion ([1,2]): These invoke diffusion, electrostatics, mechanical interlocking, or adsorption processes. The adsorptive theory is the one that best applies to the carbon fiber/thermoplastic physics. According to the model, for maximum adhesion to occur, the adhesive must come into intimate contact with the substrate. One can then use thermodynamic principles to define the work of adhesion, W_A, as the work required to separate a unit area of the two materials in order to create two new surfaces. Fowkes ([3]) has championed the importance of acid/base interactions in adhesion and in W_A. He postulated that the work of adhesion, and therefore the final performance of an adhesive bond, is dominated only by acid/base and dispersive energies. Thus, it becomes desirable to quantify these dispersive and non-dispersive (acid/base) interactions as a means

---

[1]Address correspondence to this author.

0097–6156/89/0391–0217$06.00/0

of investigating and predicting the strength of an interface. Overall, a satisfactory final performance of a composite created from fibers and thermoplastics must in some way, at least partially, reflect W_A and, therefore, a favorable matching of the two types of contributing intermolecular energies. By lumping various dipolar, hydrogen bonding and other types of possible specific interactions into the acid/base category, a great simplification results and numerical results emerge.

The technique of inverse gas chromatography (IGC) has been previously used to study the nature of carbon fiber surfaces (4); gratefully, the current authors acknowledge the important earlier investigations and followed their techniques (See also Schultz, et. al. in this book). X-ray Photoelectron Spectroscopy (XPS) was used to elucidate confirming information on the chemical composition on the carbon fiber surfaces (5). However, new methodology had to be developed to study the surface energetics of thermoplastic polymers which involved using capillary column inverse gas chromatography (CIGC). Once the surface energies and surface chemical components of various fibers and matrices were investigated, the adhesion of these fibers to the thermoplastic resins were evaluated. These data were obtained by performing a fiber critical length test (6). A correlation between dispersive and acid/base properties of the fibers and the quality of the fiber matrix adhesion was then possible.

Materials and Methods

Materials. Several precursor materials exist for the production of carbon fibers (7). However, most of the presently available carbon fibers are synthesized from polyacrylonitrile (PAN) since these fibers have the best mechanical properties. Five PAN based carbon fibers were used in this study:

1. AU-4, untreated fiber from Hercules,
2. AS-4, surface treated fiber from Hercules,
3. XAS, surface treated fiber from Dexter Hysol,
4. AU-4 treated with Z-6040, and
5. AS-4 treated with Z-6040.

The Z-6040 is a Dow Corning silane coupling agent, 3-glycidoxypropyltrimethoxysilane:

$$(CH_3O)_3SiCH_2CH_2CH_2OCH_2CH\underset{O}{\diagdown\diagup}CH_2$$

The Z-6040 treatment procedure is outlined below. AS-4 and XAS fibers were commercial samples having a proprietary surface treatment which was not available for this study. These fibers were not coated.

The thermoplastic resins that were used for the adhesion studies are considered to be tough, not easily crystallizable polymers. They are as follows, with designated labels assigned to each structure:

Tg=190°C, obtained from Scientific Polymer Products, referred to as
"polysulfone";

Tg=150°C, obtained from Scientific Polymer Products, referred to as
"polycarbonate":

Tg=210°C, obtained from General Electric (ULTEM R1000), identified
below as polyetherimide.

The organic molecules or "probes" used to investigate the
<u>dispersive</u> surface energies of the fiber surfaces were a series of
n-alkanes. The probes used to study the <u>non-dispersive</u> forces were
chosen based on their acidic or basic character as determined by
Gutmann (<u>8</u>). Gutmann has practically defined basicity as the donor
number, DN, or electron-donor capability in the Lewis sense. The
donor scale is based on the value of the molar enthalpy for the
reaction of the electron donor with a reference acceptor, $SbCl_5$.
On the other hand, the acceptor number, AN, characterizes the
acidity or electron acceptor capability of a material. It is based
on the NMR chemical shift of ^{31}P contained in $(C_2H_5)_3PO$ when
reacting with the acceptor. Each probe selected had a known AN and
DN in order to quantitatively "sample" the respective surfaces
involved in the composite. Three probes were used to study the
fiber surfaces. Chloroform ($CHCl_3$) was used as the acidic probe
and had an AN equal to 23.1 and DN equal to 0. Tetrahydrofuran
(THF) was used as the basic probe with AN equal to 8.0 and DN equal
to 20.0. Ethyl acetate (EA) is considered to be amphoteric with an
AN equal to 9.3 and DN equal to 17.1. For the fiber investigations
$CHCl_3$, THF and EA proved to be satisfactory from a chromatographic
standpoint.

The thermoplastics were investigated by CIGC well below their
glass transitions so that surface thermodynamics were dominant.
Column temperatures were varied over a 40C range in order to find
enthalpies and entropies of interaction. When injected into the
capillary columns coated with the thermoplastic polymers, the three
probes listed above exhibited peaks which tailed continuously even
at the lowest possible concentrations. For this reason, weaker
acids and bases which eluted as symmetric peaks were used to study
the polymer surfaces. Specifically, methylene chloride (CH_2Cl_2)
with an AN of 20.4 and its DN is equal to 0, and nitromethane (NO_2-
CH_3) having an AN equal to 20.5 and DN equal to 2.7 were chosen.
These chemicals were gold-label grade from Aldrich Chemical and
stored over 4 A molecular sieves before use.

<u>Pretreatment with Z-6040.</u> A dilute aqueous solution (0.5 wt%
silane concentration) was prepared. The pH of the water was

adjusted to 3.0 to 4.5 with 0.1% acetic acid, and the silane was
then added. The carbon fibers were dipped in this solution for
approximately 5 min. and then dried at 115°C to remove traces of
methanol that resulted from the hydrolysis of the methoxysilane.

X-ray Photoelectron Spectroscopy. XPS spectra of the carbon fibers
were recorded on a Perkin Elmer PHI 5300 electron spectrophotometer
with a magnesium Kα source operating at 250 mW. The operating
pressure was 4 x 10^{-7} torr. The samples were prepared by mounting
the fibers across gaps in metal holders.

Fiber Critical Length Experiment. The aluminum coupon fiber
critical length test (9) was used to obtain the critical length
data. Coupons of annealed Al100 aluminum measuring 2.5 x 15.2 cm
were prepared by wet sanding with 400 grit sand paper. The coupons
were coated with approximately 3 mL of a 5 wt% polymer solution in
methylene chloride solvent. The solvent was allowed to evaporate
at room temperature for 24 hours. Single fibers (3 or 4 per
coupon) were placed on the polymer film parallel to the long axis
of the coupon. The fibers were then coated with another 3 mL of
the polymer solution and again the solvent was allowed to
evaporate. The coupons were annealed at 10°C above the Tg of the
polymer followed by 265°C for 8 hours. After cooling, the samples
were placed in an Instron testing machine and pulled in tension to
30% strain at 25% per minute strain rate. The lengths of broken
fibers were measured on a microscope with a micrometer stage.

Chromatographic Conditions. IGC measurements were carried out
using a Hewlett-Packard 5890 gas chromatograph equipped with a
flame ionization detector. A 1.0 m stainless steel column with an
internal diameter of 4.4 mm was packed with 8 to 9 g of carbon
fiber by pulling approximately ten 1.0 m long tows of fiber through
the column. Helium was used as the carrier gas and methane as the
non-interacting marker. The flow rate was 13.5 mL/minute. The
injector temperature was 200°C and the detector temperature was
250°C. The CIGC was performed with 60 m fused silica columns with
an internal diameter of 0.53 mm. Columns, supplied by Hewlett-
Packard Avondale Division, were statically coated (10) with a
resulting film thickness of 2 to 3 μm. The carrier gas was
hydrogen at a flow rate of 5 to 10 mL/minute. This analysis was
automated by the use of a Hewlett-Packard 19395A headspace sampler.
All columns were conditioned overnight at 110°C prior to use. For
greater detail about the experimental techniques and procedures,
see the chapter by Bolvari, Ward, Koning, and Sheehy in this book.

Inverse Gas Chromatography. The IGC results followed from
measuring the retention times of the probe molecules injected into
the columns packed with the fiber or coated with the polymer. To
measure the dispersive interactions, the non-polar n-alkane probes
were used. For the acid/base (or non-dispersive) interactions of
the fibers, $CHCl_3$, THF, and EA were used. On the other hand,
CH_2Cl_2 and nitromethane were the nondispersive probes for the
thermoplastics for reasons discussed above.

The net retention volume, V_N, was calculated from ([11]):

$$V_N = jD(t_r - t_o),\qquad(1)$$

where t_r is the retention time of the probe, t_o is the retention time of the non-interacting marker (methane), D is the flow rate, and j is a correction factor for gas compressibility.

The following relationship was used to calculate the dispersive component of the surface free energy, γ_S^D, for the fibers ([4]):

$$RT\ln V_N = 2\,N\,(\gamma_S^D)^{\frac{1}{2}} a (\gamma_L^D)^{\frac{1}{2}},\qquad(2)$$

where N is Avogadro's number, a is the surface area of the probe molecule, and γ_L^D is the dispersive component of the surface free energy of the liquid (probe). The slope of the reference n-alkane line in a plot of $RT\ln V_N$ versus $a(\gamma_L^D)^{1/2}$, is equal to $2\,N\,(\gamma_S^D)^{1/2}$. The specific free energy corresponding to the specific acid/base interaction, $\Delta G^\circ sp$, was calculated using ([4]):

$$\Delta G_{sp}^\circ = RT\ln\,(V_N/V_N^{ref})\qquad(3)$$

From a graphical point of view, equation 3 reveals that data points from probes that fall above the n-alkane line do so because of non-dispersive or acid/base interactions. ΔG_{sp}° was obtained from a ratio of the data point of the acidic or basic probe to that on the reference line at a given value of $a(\gamma_L^D)^{1/2}$. The specific enthalpy, ΔH_{sp}°, and specific entropy, $\Delta S^\circ sp$, were obtained by analyzing the IGC experiment at various temperatures. A plot of ΔG_{sp}° versus T had a slope of ΔS_{sp}° and an intercept of ΔH_{sp}°, as indicated by:

$$\Delta G_{sp}^\circ = \Delta H_{sp}^\circ - T\Delta S_{sp}^\circ\qquad(4)$$

Results and Discussion

X-ray Photoelectron Spectroscopy (XPS).

In XPS, the sample is bombarded with soft x-rays and the photoelectrons emitted are analyzed in terms of their kinetic energy, E_K. The resulting core level peaks such as the C 1s are due to photoelectrons emitted from the atomic (core) orbitals of the atoms in the surface layers present. The binding energies, E_B, of these electrons were obtained from:

$$E_B = h\gamma - E_K - \phi,\qquad(5)$$

where $h\gamma$ is the x-ray energy and ϕ is the sample work function ([12]). The binding energies are highly characteristic and allow identification of all elements except hydrogen. The peak intensities at these characteristic binding energies are proportional to the number of atoms sampled and atomic compositions can therefore by calculated. The XPS data shown in Table I which lists the concentration in atomic % and binding energies of four

Table I. Atomic Percentages (%) and Binding Energies (BE)
in ev of Carbon, Oxygen, Nitrogen, and Silicon

	C 1s		O 1s		N 1s		Si 2p	
	%	BE	%	BE	%	BE	%	BE
AU-4	89.58	285.0	8.46	533.1	1.96	400.2	---	-----
AS-4	79.66	285.0	14.52	533.1	5.82	400.5	---	-----
XAS	76.92	284.9	15.95	532.4	7.13	400.1	---	-----
AU-4+ Z-6040	77.19	285.6	18.03	533.1	1.25	400.8	3.52	102.8
AS-4+ Z-6040	63.88	285.2	27.64	533.0	2.29	400.3	6.19	102.6

elements of interest. The commercially treated fibers (XAS and AS-
4) have almost double the oxygen content of the untreated AU-4
fiber in their upper 50 A° of surface. This indicates that the XAS
and AS-4 fibers have oxygen containing functionalities which were
introduced as a result of surface pretreatment. Thus, these fibers
have groups on their surface which are potentially able to interact
with the polymer matrix. More information concerning the acidity
or basicity of these functional groups were obtained by IGC (see
below).

Two fibers were treated with the silane coupling agent. The
AS-4 fiber treated with Z-6040 shows a higher concentration of Si
and O than does the untreated AU-4 fiber as shown in Table I.
Clearly, more of the coupling agent has been added to the AS-4
fiber. Although a silane coupling agent would normally form a Si-
O-Si bond with a Si-OH containing glass surface, here the bond that
is most likely to be formed is C-O-Si. The Z-6040 treated AS-4
fiber, which has more oxygen containing functionality on its
surface, would logically form more of these bonds, resulting in a
higher concentration of O and Si. All four treated fibers (AS-4,
XAS, and the two fibers treated with Z-6040) are expected to have
greater potential for non-dispersive interactions based on their
higher content of oxygen and nitrogen when compared to the
untreated AU-4 fiber.

Dispersive Interactions. IGC was used to obtain the dispersive
component of the surface free energy of the carbon fibers and of
the polymers (Equation 2). These results are shown in Table II.
The dispersive component reflects the graphitic nature of the
fiber. Indeed, a more graphitic structure has more conduction
electrons; and, thus, has more polarizable electrons to contribute
to the dispersive force interaction. Pure graphite has a
dispersive surface energy component of 150 mJ/m^2 ($\underline{13}$). It should
be noted that the AU-4 fiber has the highest γ_D, with the surface
treated AS-4 and XAS having slightly lower γ_S^{DS} values. Therefore
the surface treatment seems to decrease the graphitic nature of the

Table II. Summary of Dispersive Components of Carbon Fibers and
Thermoplastic Polymers Obtained by IGC

Fibers	$\gamma_s^D (mJ/m^2)$	Polymers	$\gamma_s^D (mJ/m^2)$
AU-4	65.1	Polysulfone	42.8
AS-4	47.5	Polycarbonate	36.2
XAS	39.3	Polyetherimide	45.5
AU-4+ Z-6040	26.6		
AU-4+ Z-6040	23.3		

fiber. The treatment with the silane coupling agent lowers the
dispersive component of surface energy considerably.

Non-dispersive Interactions. Figure la and lb are illustrative
plots of $RTlnV_N$ versus $a(\gamma_L^D)^{1/2}$ for the AS-4 fiber and the
polysulfone. The n-alkane data based line has a slope related to
γ_s^D for the material. The points corresponding to the acidic,
basic, and amphoteric probes clearly lie above the n-alkane
reference line indicating that non-dispersive interactions are
coming into play. The specific free energy was calculated for
these probes using Equation 3. Specific free energy as a function
of temperature for AS-4 is shown in Figure 2. The specific
enthalpies and entropies were also calculated. These heats or
enthalpies of interaction are shown in Table III. The specific
enthalpy was negative in all cases, indicating a favorable,
exothermic, interaction. For each fiber the heat of interaction
was greater for the basic probe (THF) and amphoteric probe (EA)
than for the acidic probe ($CHCl_3$). Although $CHCl_3$ and THF do not
have the same relative acid and base strengths ($CHCl_3$ is a stronger
acid than THF is a base) the interactions were still less favorable
for the acid indicating that the fiber surface of AS-4 is
predominantly acidic. The heat of interaction for the treated
fibers XAS and AS-4 were greater than for the untreated fiber AU-4.
Thus, these proprietary surface treatments must have resulted in
the addition of functional groups potentially able to interact with
the matrix. The XAS fiber was slightly more basic in nature than
the AS-4 fiber, since its interaction with $CHCl_3$ was greater and
its interaction with THF was less.

In the case of the fibers treated with the silane coupling
agent, there was a dramatic lowering in the heats of interaction of
all probes. Since the Z-6040 coupling agent has introduced the
basic epoxy group, it follows that the acidic probe had a heat of
interaction comparable to that of the basic probe (THF); that is,
the specific free enthalpy gap between acidic and basic interaction
was narrowed by the coupling agent.

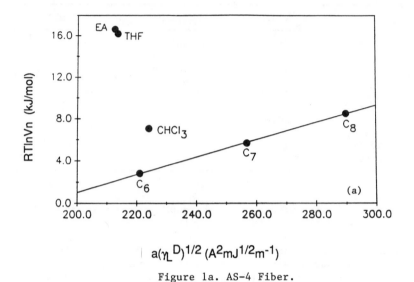

Figure 1a. AS-4 Fiber.

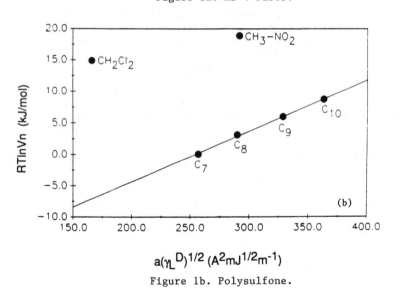

Figure 1b. Polysulfone.

Table III.   Summary of Specific Heats for Fibers and Polymers

Material	$-\Delta H^{\circ}_{sp}$ (kJ/mol)		
	EA	THF	$CHCl_3$
AU-4	47.7	45.7	35.8
AS-4	63.8	62.4	44.9
XAS	60.9	58.2	46.3
		CH_2Cl_2	NO_2-CH_3
Polycarbonate		57.0	54.6
Polysulfone		60.8	54.7

Overall the polymers investigated were predominantly basic since they interacted most strongly with the acidic probes. Nitromethane, which is slightly basic but has the same acid strength as methylene chloride, had an enthalpy of interaction lower than that for the methylene chloride. This indicated that the basicity of nitromethane resulted in a less favorable interaction with the basic polymer surfaces. The non-dispersive interactions were not determined for polyetherimide since all of these probes resulted in nonsymetric, tailing peaks. Perhaps the polyetherimide is quite a strong base and attracted the acidic probes sufficiently to prevent equilibrium thermodynamic measurements.

Fiber–Matrix Adhesion. When the aluminum specimen described in the Materials and Method section was pulled in tension, the fiber, which is oriented axially within the test coupon, breaks as shown in Figure 3. Under tensile loading, shear forces are transferred from the matrix to the fiber at the interface. The transfer causes a build-up of tensile forces in the fiber until the local tensile strength of the fiber is exceeded. The fiber then fractures within the polymer. This process accumulates until the fragments remaining are no longer large enough to support sufficient shearing forces to exceed the fiber tensile strength. The average length of the broken fiber fragments is referred to as the fiber critical length (Lc). The Lc is an indication of the ability of the polymeric matrix to transfer stress to the fiber; therefore, it is also an indication of the quality of the fiber-matrix adhesion, smaller Lc meaning stronger interfacial strength.

The critical fiber length for the fibers in the three different polymer matrices are shown in Table IV. Based on these critical lengths, both commercially treated fibers (XAS and AS-4) showed better adhesion to the polymer matrices than did the untreated fiber (AU-4). This can be explained in terms of the greater acid/base interactions of these two fibers with the matrices.

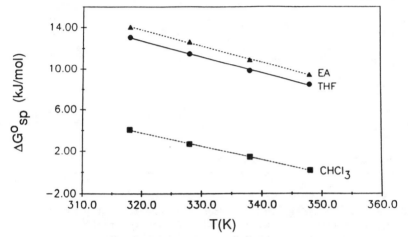

Figure 2. Specific Free Energy Versus
Temperature for AS-4 Fiber.

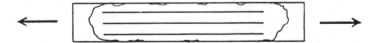

Figure 3.   Schematic Diagram of a Fiber Critical Length
Experiment Using an Aluminum Substrate.

Table IV. Fiber Critical Lengths (Lc, mm) and Standard Deviations
(SD) of Fibers Embedded in Polysulfone (PS), Polycarbonate
(PC) and Polyetherimide (PEI)

	POLYMER					
	PS		PC		PEI	
	Lc	S.D.	Lc	S.D.	Lc	S.D.
FIBER						
AU-4	0.59	0.18	0.65	0.20	0.68	0.21
AS-4	0.44	0.16	0.46	0.20	0.26	0.06
XAS	0.33	0.10	0.26	0.06	0.25	0.05
AU-4+ Z-6040	0.46	0.28	0.45	0.30	0.46	0.32
AS-4+ Z-6040	0.39	0.24	0.36	0.16	0.27	0.11

The XAS fiber was found to have the best adhesion to the
polymers. This is difficult to rationalize on the exclusive basis
of acid/base interactions. On the other hand, recall that the AS-4
fiber was determined to have a more graphitic structure than the
XAS fiber (this was also confirmed by the γ_s^D measurement). Bascom
(15) and coworkers have suggested that highly graphitized surfaces
(large dispersion component of surface energy) may bind water
tightly. This water may not be fully stripped in the column
conditioning process and would lead to lower interfacial shear
strengths. Thus, the least graphitic XAS fiber may overcome its
slightly lower predicted enthalpy of interaction (compared to AS-4)
with the thermoplastic matrices and adhere quite well. The fiber
structure may be a factor in the quality of the interface as well.
This will be investigated in future experiments.
The fibers treated with Z-6040 showed some interesting
results. When viewing under the microscope, two modes of failure
were observed at different areas across the fiber (14); these are
frictional stress transfer (interface unbonding) depicted below,

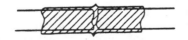

and shear stress transfer shown as follows:

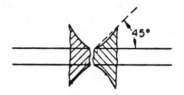

In order to interpret the two types of failure, it is reasonable
that interface debonding predominates when the fiber matrix
adhesion is poor. In contrast, if a ductile matrix is used and the
adhesion is high, the matrix will fail more by the shear dominated
process. For several millimeters along the treated fiber shear
bands were observed and the Lc was small (comparable to the XAS
fiber). However, further along the fiber, interface unbonding was
seen. At these areas, the Lc was large (comparable to the AU-4
fiber). This is the reason for the large standard deviation seen
in Table IV for the silane modified fibers. This could be related
to poor coating technique; that is, the fiber was not completely in
contact with the coupling agent. Also the coupling agent may have
failed to bond to the fiber even though they were in contact. A
third possibility is that the C-O-Si bond was formed but was
unstable under the testing conditions.

Conclusions

The nature of carbon fiber surfaces were investigated by IGC in
terms of dispersive and non-dispersive contributions to total
surface energy. The acidic or basic properties of the
thermoplastic polymers was quantified by CIGC.
     Single fiber tests were used to evaluate the fiber-matrix
adhesion. Acid/base interactions do play a role in this adhesion
but are not exclusively responsible for improvements at the
interface. Treatment with a siloxane coupling agent was proven to
be effective in lowering the dispersive component of the carbon
fibers and altering their acid/base properties, producing poorer
adhesion.

Literature Cited

1. Wake, W. D., Adhesion and the Formulation of Adhesives, 2nd
    Ed., Applied Science Publishers, 1982.
2. Kinloch, A. J., Durability of Structural Adhesives, Applied
    Science Publishers, 1982.
3. Fowkes, F. M., J. Phys. Chem., 1962, 66, 382.
4. Schultz, J.; Lavielle, L.; Martin, C., J. Adhesion, 1987, 23,
    45.
5. Ishitani, A., Carbon, 1981, 19, 269.
6. Drzal, L. T.; Rich, M. J.; Camping, J. D.; Park, W. J., Proc.
    35th Ann. Tech Conf. Reinf. Plastics/Composites Inst. SPI 20-
    C, 1980, pp. 1-7.
7. Donnet, J. B.; Cazeneuve, C.; Schultz, J.; Shanahan, M.E.R.,
    Proc. Int. Conf. Adhesion and Adhesives, 1980, 19, 1.

8. Gutmann, V. The Donor Acceptor Approach to Molecular Interactions, Plenum Press, 1978.
9. Wadsworth, N. J.; Spilling, I., Brit. J. Appl. Phys., 1968, 1, 1049.
10. Jennings, W., Gas Chromatography with Glass Capillary Columns, 2nd Ed., Academic Press, 1980.
11. Lipson, J. E.; Guillet, J. E., in Developments in Polymer Characterization-3 Ed.; Dawkins, J. V., 1982, pp. 33-74.
12. Brewis, D. M.; Briggs, D., Industrial Adhesion Problems, Orbital Press, 1985.
13. Donnet, J. B., Carbon, 1982, 20, 267.
14. Mullin, J. V.; Mazzio, V. F., J. Mech. Phys. Solids, 1972, 20, 391.
15. Bascom, W. D.; Yon, K-J.; Jensen, R. M.; Cordner, L, Presnetation at "A Multidisciplinary Workshop on Chemistry and Properties of High Performance Composites", Jackson, Wyoming (1988).

RECEIVED September 29, 1988

Chapter 17

Surface Characteristics of Glass Fibers

E. Osmont and Henry P. Schreiber

Chemical Engineering Department, Ecole Polytechnique, Montréal, Québec
H3C 3A7, Canada

Inverse Gas Chromatography (IGC) was applied to
E-glass fiber surfaces modified by various
silane coupling agents. Using homologous series
of alcohol (acid) and amine (base) vapor
probes, the acid/base interaction characteristics of the fibers were measured from 30 to
90°C. Unmodified E-glass was found to be
amphipatic, significantly bonding with both
acid and base vapors. Strong surface acidity
was produced by a chloro-silane agent (CPTMS),
while surface treatments with hexyldimethoxysilane (HDMS) and an aminosilane (APS) generated
increasing degrees of basicity. The temperature
dependence of these surface properties was
established. Adsorption from solution of
polyvinyl chloride (PVC) and of PMMA, respectively known to be an acid and a base from IGC
measurements, showed the existence of selective
sorption effects. PVC was strongly sorbed on
basic glasses, while PMMA sorbed more readily
on the acidic, CPTMS-treated glass. These findings are an origin for more extensive studies of the role played by interfaces in
composites reinforced by the various fibers.

Since its introduction some years ago, inverse gas chromatography
(IGC) has been recognized as a convenient route to the determination of thermodynamic interaction parameters for polymeric or
other non-volatile stationary phases in contact with selected
vapor probes ($\underline{1},\underline{2}$). The principles of IGC experiments have also
been extended to two-component stationary phases ($\underline{3}$), thereby
making it possible to specify thermodynamic interaction parameters
for the components of polymer blends ($\underline{4},\underline{5}$), as well as for filled
polymers and other multi-component systems. Despite these
attractive features, limitations must by recognized on the general

0097–6156/89/0391–0230$06.00/0
© 1989 American Chemical Society

applicability of IGC to the measurement of interaction thermodynamics (6). This is due in part, in some cases, to the preferential partitioning of the probe molecule to one of the components in a mixed stationary phase. Further, the calculation of thermodynamic functions depends on the existence of equilibria between volatile and stationary phases. Failure to attain equilibrium compromises the validity of thermodynamic computations, even when only a single stationary phase is present. Equilibrium conditions are readily attained when only non-polar materials are involved, but tend to break down when the materials used in the IGC experiment can interact by non-dispersive forces.

The above limitations are particularly restricting to the many applications of multi-component systems in which highly polar materials are used and therefore prompted recent modifications of the IGC method (7,8). These modifications result in the generation of comparative, internally consistent indexes of acid or base functionality for a wide range of polymers and the constituents for polymer systems. Although the formality of thermodynamic functions is lost, the acid/base interaction indexes promise to be useful in rationalizing such important aspects of behavior as the dispersion of particulates in polymeric fluids, the development of mechanical properties in composites, and the diffusion of vapors through polymeric membranes (9). The present paper considers the acid/base interaction potential of glass fibers. Of particular interest is the range of these interaction potentials that can be designed through surface modification of fibers by various silane coupling agents. The consequences of diverse acid/base potentials are illustrated by the adsorption onto the fibers of polymers known to be acidic or basic. The work adds to the recent literature (10,11) on the use of IGC for the surface characterization of glass fibers.

Experimental Section.
i.   Materials.   Four glass fiber specimens were used. One was an unsized E-glass, the others were surface treated with silane coupling agents as follows:
CPTMS (3-chloropropyltrimethoxy silane) was applied from isopropanol/water mixtures at pH 4 (acidification by glacial acetic acid).
HDMS (Hexyldimethylethoxy silane) was applied from acetone solution.
APS (4-aminobutyldimethylethoxy silane) was also applied from acetone solutions.

For ease of packing chromatographic columns, the fibers were screened, with 325 to 400 mesh particles retained for the study. Columns 1.5 m long were constructed of stainless steel tubing that had been degreased, washed, and dried. The columns were used for IGC work with a Perkin-Elmer Sigma II chromatograph, fitted with dual flame ionization detectors. The vapor probes were reagent grade n-octane, ethanol, n-propanol, n-butanol, propylamine, butylamine, and ethylene diamine. The octane was used as a probe capable of interacting with substrates through van der Waals forces only. The alcohols represented Lewis acids and the amines represented Lewis bases (12,13).

In adsorption measurements, PMMA and PVC were the adsorbing molecules, the former from solutions in toluene, the latter from THF solutions. Earlier work had shown PVC to be a strong acid (7,8), while Fowkes (14) reports PMMA to be a Lewis base. The polymers were commercial samples, the PMMA with Mw = 1.13 x10^5 and the PVC with Mw about 6.5 x 10^4. Adsorption measurements were uniformly at 30 ± 1°C.

ii Procedures In IGC determinations, columns were swept with dry nitrogen at 140°C for approximately 1h prior to the introduction of vapor probes. Vapors were injected at extreme dilution by microsyringes, and experimental temperatures ranged from 30 to 90°C. Inlet pressures were in the vicinity of 20 psig. and He carrier gas flow rates were controlled at 15 ± 1 ml/min. Octane probes generated symmetrical elution peaks, leading to standard calculations of retention times and specific retention volumes, V_g^0 (1,2). Polar probes generated skewed peaks, necessitating measurement conventions described in detail in Reference 7. All data quoted here are averages of 5 separate vapor injections. The V_g^0 data bear the following errors: For n-octane, ± 2% over the entire temperature range; for polar probes, ± 5% at T<50°C, and ± 8% at T>50. The increased high temperature uncertainty is due to the diminished values of V_g^0 at higher experimental temperatures.

For adsorption studies, a polymer solution was prepared at an initial solute concentration in the span 0.5 to 2.5 wt-%. A carefully weighed sample of glass fiber was introduced into a 100-ml or 250-ml aliquot and shaken for 24 h at the experimental temperature. Following an additional 24 to 48 h. period for sedimentation, the supernatant fluid was filtered through a coarse glass plug. An aliquot of the clear liquid was evaporated to dryness under vacuum at 60°C. Pumping was continued for several hours following the attainment of invariant solids weight, using a Sartorius microbalance for the latter purpose. Polymer adsorbed, Cads., was calculated from the difference in initial and final solution concentrations and expressed as weight adsorbed/unit area of glass surface. For this purpose, an apparent surface area of 0.22 m2/gm was used, based on microscopic evaluation of fiber geometry. The calculation assumes that the fibers were non-porous and that samples viewed by microscopy were representative of the bulk. Data reproducibility was from ± 7 to ± 9% in all cases.

Results and Discussion
i. IGC Results Table I summarizes the composition of columns used in this work. The relatively high solid loadings were necessitated by the low retention volumes resulting from low specific surface areas. Specific retention volumes and acid/base interaction parameters are given in Table II; all data are referred to a reference temperature of 30°C. The interaction parameter, Ω, was calculated from the retention volumes for n-butanol and butylamine, following the precedents of References 7

and 8. Accordingly, for acidic substrates, where the $[V_g^0]_{base}$ exceeds that for the acidic alcohol,

$$\Omega = 1 - (V_g^0)_b / (V_g^0)_a < 0 \qquad (1)$$

For basic stationary phases, where the $[V_g^0]_{acid}$ exceeds that for the basic butylamine,

$$\Omega = (V_g^0)_a / (V_g^0)_b - 1 > 0 \qquad (2)$$

On this basis, unsized E-glass is a mild acid, its Ω value falling outside the band of values near 0, that is generally associated with amphipatic solids or with materials able to interact through dispersion forces alone. The effects of surface treatment are quite distinct with appreciable acidity being introduced by CPTMS coatings (column 2), and moderate basicity by APS (column 4). HMDS sizing produces relatively mild changes in surface condition, the net effect being a weak surface basicity. Table II reports the Ω values for PVC and PMMA, as determined earlier. The former is a strong acid and the latter is distinctly basic, in agreement with the findings of Fowkes and coworkers (14).

Table I. Column Description for IGC Experiments

Column Number:	1	2	3	4
Stationary Phase	E-glass	E-glass	E-glass	E-glass
Surface Treatment	Nil	CPTMS	HDMS	APS
Wt. of stationary phase (g)	5.27	4.43	5.06	5.82

CPTMS = 3 - chloropropyltrimethoxysilane
HDMS = hexyldimethoxysilane
APS = aminopropyltrimethoxysilane

The low V_g^0 in Table II warrant comment. Pristine glass is known to have highly reactive surfaces (15), so that large retention volumes might be expected. However, as shown by Shafrin and Zisman (15), surface activity in aged glass surfaces is greatly reduced through the presence, often through chemical adsorption, of layers able to produce either acid or base surface characteristics. The preconditioning in our experiments was insufficient to free the surfaces of strongly bonded species; therefore, the data for the unsized surfaces cannot be ascribed to the properties of truly bare glass. Another contribution to the low V_g^0 values arises from the convention of computing chromatographic results on the basis of weight of stationary phase. When solids with low surface areas are involved, this leads to an

Table II.　Comparison of Retention Volumes at 30°C
(Retention volumes in ml. g^{-1})

Column Number:	1	2		3		4	
Probe:							
nC8	1.77	2.46		1.81		3.11	
Butanol	6.08	5.25		4.07		7.44	
		>0.8	>2.1		>0.7		>0.5
Propanol	5.27	3.14		3.25		6.91	
		>1.1	>0.5		>0.5		>0.6
Ethanol	4.06	2.58		2.69		6.34	
Butylamine	6.93	17.48		3.51		4.07	
		>1.2	>1.3		>1.2		>1.3
Propylamine	5.74	16.14		2.26		2.69	
Ethylene diamine	8.83	17.51		3.59		2.26	
Ω^*	-0.14	-2.33		0.16		0.83	

* Calculated from V_g^0 for butanol and butyl amine.

For polymers used in this work; $\Omega_{PVC} = -1.33$; $\Omega_{PMMA} = 0.91$, both at 30°C.

apparent reduction in V_g^0, which is somewhat misleading. For example, recent work with rutile (TiO_2) stationary phases ([7]), has given octane retention volumes in the vicinity of 100 $ml.g^{-1}$. Specific surfaces in these cases were approximately 8 m^2/g, which is about 30 times greater than in the present work. By recalculating the present data on a basis of unit surface area, it leads to V_g^0 in the range of 50 to 100 $ml.g^{-1}$, and not greatly out of line with other particulate stationary phases used in IGC experiments.

A final comment in connection with Table II pertains to the probe-to-probe variation in V_g^0. A general decrease in V_g^0 with decreasing probe chain length is indicated by the offset numbers in Table II. On a percent basis, the variation is small when acid/base interactions occur between probe/substrate pairs, but becomes considerably larger when acid/acid or base/base pairs are in contact. At constant T, the increasing volatility of shorter-chain probes may be expected to produce variations, such as those exemplified by (-OH) probes in column 2, and by (-NH$_2$) probes in column 4. That is, changes of 20 to 40% per CH_2-group are observed, when only dispersion forces are active. The variations of V_g^0 per amino group in column 2 and per hydroxyl group in column 4, amounting to less than 10% per CH_2 group, suggest that the strength of acid/base interactions decreases with increasing alkyl chain length. The increasing importance of weaker, dispersion force contributions to the overall bonding energy may be responsible for the observation. Finally, though the retention volume for the diamine probe in contact with CPTMS-treated glass is appreciable, it is in line with values for other basic probes used in this work. The result, indicative of an apparent head-to-tail orientation of the probe molecule on the glass surface, displays another facet of the applications to which the IGC method may be put.

The temperature dependence of interaction variables for components of reinforced polymer systems is of great importance, given the range of temperatures over which such systems are processed and used. Since the thermodynamic basis of IGC parameters links these with interaction enthalpies ([1],[2]), interaction data originating from IGC will follow the trends set by the formal thermodynamic data. In this regard, the IGC method has a considerable advantage over alternative methods for evaluating interactions among the components of non-volatile materials. Since IGC experiments are readily carried out over wide temperature ranges, they overcome difficulties inherent in the use of other parameters, such as the solubility parameter, which is generally estimated at fixed temperatures, or over narrow ranges of the variable, leaving undetected possible changes in the miscibility of components. The temperature scan in this work was limited primarily by increasing experimental uncertainties as retention times decreased with rising temperature. Nevertheless, useful results were obtained to a maximum of 90°C. These results

are displayed in Figures 1,2, and 3 in terms of Arrhenius-type
plots showing In V_g^0 as a function of reciprocal absolute
temperature. Figure 1 gives the temperature dependence for the
octane probe, Figure 2 gives the acidic butanol, and Figure 3
gives the results for the butylamine probe.
    The representation in Figure 1 leads to linear and roughly
parallel relationships in all cases. Thus, although the absolute
retention capacities of the glass substrates vary with the
applied surface sizing, the bonding energies with n-octane are
constant. In other words, the number of interaction sites for
octane may be considered greater on APS- and CPTMS- treated fibers
than on HMDS and untreated versions of the fiber; however, the
forces involved at the probe/substrate contact are the same. This
is not unexpected, given the non-polar nature of the probe. The
pattern of results in Figures 2 and 3 is quite different. When
acidic butanol is the probe molecule, linear relations are
generated with the strongly acidic CPTMS-sized glass substrate,
but increasing non-linearity is observed when going in the
direction of increased substrate basicity, that is, in the
sequence HMDS, unsized, and APS-treated glass. Conversely, when
using the amine probe, non-linearity is produced with the acidic,
CPTMS-sized glass, and to a lesser degree with unsized glass, but
essentially linear plots are obtained with the basic surfaces.
    The systematic differences in temperature dependence discussed
above are related to the presence or absence of acid/base
functionality at interfacial contacts. Inspection of Figures 1, 2
and 3 shows that the predominance of dispersion forces results in
what seem to be roughly constant slopes in all cases. Significant
slope reductions are noted in Figures 2 and 3 for those cases
where acid/base forces are expected to be significant factors. In
those instances, (for example butanol/APS-treated glass in Figure
2 or butylamine/CPTMS-treated glass in Figure 3) there are shifts
from lesser to greater slopes as temperatures rise into the 50   to
-70°C range. At the same time, a pronounced reduction in the V_g^0
values is apparent. A tentative interpretation calls for
molecular crowding of those probe vapors capable of interacting
with the substrates through non-dispersive forces. These forces
weaken at increasing temperatures until the probe molecules,
whether polar or not, adopt surface orientations similar to those
taken on by octane. Thereupon, they interact with the surface
mainly through van der Waals forces.
    Partial justification of these speculations follow from
activation energies computed from Figures 1 to 3, and reported in
Table III. In addition Table III contains activation energies for
the retention of the diamine probe, not included in the preceding
figures. The data are calculated only for linear portions of the
Arrhenius representations. A roughly constant value near 2.5
Kcal/mol is obtained when dispersion forces are dominant. This
applies not only to the octane probe, but to acid/acid and
base/base pairs as well. The strongest acid/base contributions
are produced by -OH/APS, diamine/CPTMS, and to a lesser degree
$-NH_2$/CPTMS interactions. This follows the sequence of interaction

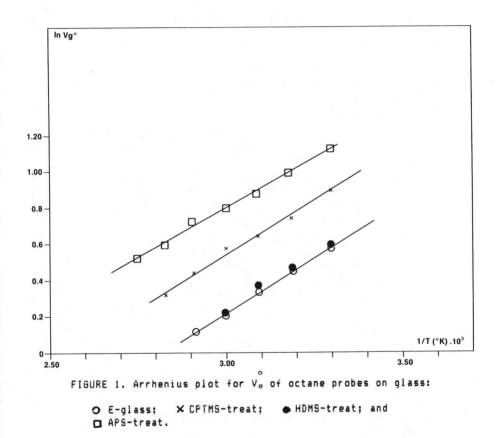

FIGURE 1. Arrhenius plot for V_g^o of octane probes on glass:

O E-glass;   × CPTMS-treat;   ● HDMS-treat; and
□ APS-treat.

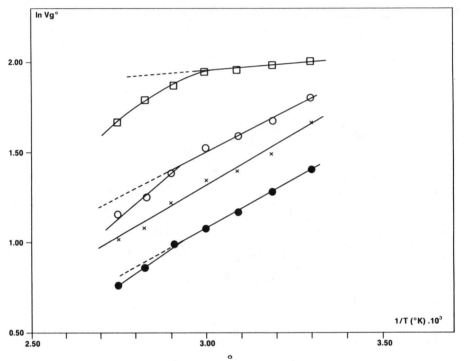

FIGURE 2. Arrhenius plot for V_g^o of butanol probes on glass:

O E-glass;   X CPTMS-treat;   ● HDMS-treat;   and
□ APS-treat.

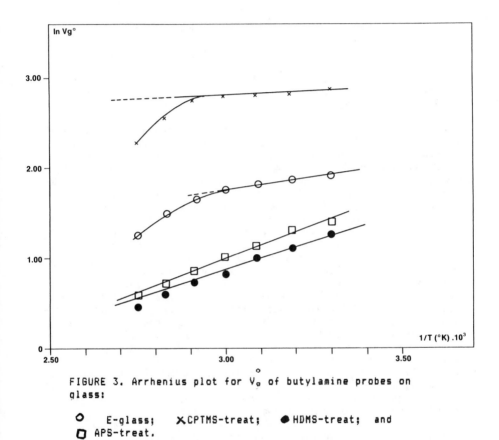

FIGURE 3. Arrhenius plot for V_g^o of butylamine probes on glass:

○   E-glass;   ✕CPTMS-treat;   ● HDMS-treat;   and
□ APS-treat.

potentials indicated by the respective Ω parameters, as shown in Figure 4.

Table III.   Activation Energies for Probe/Substrate Interactions
(E^{+} in Kcal.mol^{-1})

Substrate: Probe:	n-Octane	Butanol	Butylamine	Ethylene Diamine
E-glass	2.4	2.7	1.2	1.1
CPTMS-sized	2.8	2.6	0.44	0.23
HDMS-sized	2.0	2.3	2.1	2.0
APS-sized	2.4	0.25	2.7	3.1

A final reflection of the complex temperature dependence phenomena is seen in Figure 5, showing the variation of Ω itself. Interestingly, the pronounced acid and base characteristics produced by CPTMS and APS treatments, respectively, increase as the temperature rises above the reference 30°C, reaching a broad peak at approximately 60°C. At higher temperatures, acid and base strengths decrease. This permits an elaboration to be made on the ideas advanced above. In the cases studied, acid/base and dispersive force interactions coexist, the latter being more thermolabile at lower temperatures. As a result, inherent surface acidity or basicity increases at first with rising temperatures, and decreases substantially once a specific temperature is exceeded. This temperature overcomes the energy barrier for the detachment of molecules retained by the operative acid/base forces. An additional observation arising from Figure 5, pertains to polymer processing. The data is such that at processing temperatures, typically above approximately 150°C, all of the glass surfaces appear to be amphipatic, with Ω near 0. From these various arguments, it is reasonable to conclude that:
  - Silane coupling agents exert powerful effects on the interaction potential of glass fiber surfaces, enabling the user to design either acidic or basic functionality into these reinforcing structures for polymer composites.
  - The existence of acid/base interactions results in distinct orientations of adsorbing moieties at the substrate surface and allows for greater concentrations of sorbed species than is the case when only dispersion forces are active.
  - Strong non-dispersive interactions are temperature dependent and their importance in the cases studied here diminishes greatly above approximately 70°C.
  - Different interactions may exist between reinforcing moieties and polymer matrixes at processing, as opposed to use temperatures. General trends suggest that surface modifications exert greater influence on use properties than on processing behavior.

ii. Sorption Results: Acid/base interactions are expected to produce preferential associations among the affected components of

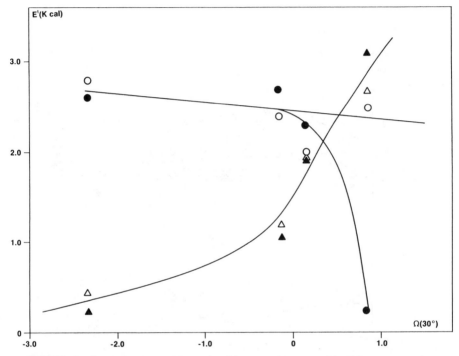

FIGURE 4. Dependence of retention volume activation energies on acid/base interactions. Probes:

   ○ octane;   ● butanol; △ butylamine; and
   ▲ diamine.

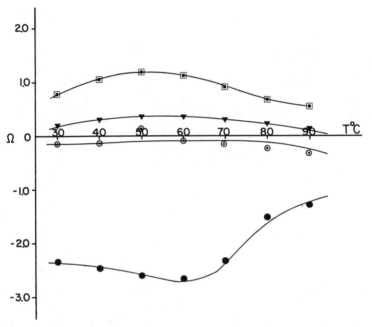

FIGURE 5. Temperature   dependence   of   interaction   parameter
for glass fibers:

⊙   E-glass;   ● CPTMS-treat;   ▼ HDMS-treat;   and
▣ APS-treat.

a polymer system. Local compositional heterogeneity is a consequence of these associations, thereby lending support to contentions that, as a general rule, multi-component polymer systems are considered as heterogeneous (8,16). Preferential adsorption, driven by acid/base forces, was demonstrated recently by Fowkes (17), for polymers interacting with filler particles. The behavior of the variously sized glass fibers as adsorbents for macromolecules selected for present purposes, is summarized in Figures 6 and 7. The former shows the sorption of acidic PVC; the latter shows the adsorption of basic PMMA.

Both sorption sequences display the expected selectivity. The adsorption of PVC is favored on APS-treated glass, while PMMA is adsorbed most voluminously on CPTMS-sized glass. The adsorption processes in both cases are affected by the solvents used. Both THF and toluene are considered to be bases (17), especially the former. Therefore, these solvents tend to compete for adsorption sites, particularly on acidic surfaces. This may inhibit somewhat the adsorption of PMMA; it may also favor the sorption of PVC on basic substrates. Further, with regard to the adsorption data, both polymers produce Langmuir-type isotherms when glass substrates are amphipatic (E-glass and HMDS-treatment), or are of like acid/base functionality as the polymer. The plateaus characteristic of Langmuir isotherms are not produced when strong acid/base interactions are implied (that is, PVC on APS-treated glass and PMMA on CPTMS-treated glass). Qualitatively, this indicates the formation of polymeric monolayers in the former cases and the development of multilayers in the latter, notably in regions of the surface marked by the presence of strong acid or base sites. Further surface diagnostics are required to elaborate on these hypotheses.

Finally, to support the qualitative contention of acid/base driving forces for the selective sorption displayed above, the adsorption data were plotted against the Ω value of the various glass substrates. The results of this procedure are shown in Figure 8. To avoid excessive crowding, Figure 8 is restricted to adsorption data from solutions with initial polymer concentrations at nominally 1.0 and 1.5 wt-%. The pattern of results is representative of the entire adsorption sequence. The strong correlation between acid/base driving forces and adsorption behavior is unmistakeable. While isotherms were conducted at 30°C only, the trends in Figure 8 should persist and become accentuated to approximately 60°C, judging from the temperature variation of Ω documented in Figure 5. At higher temperatures, the tendency should diminish and under processing conditions, no preferential sorption effects should remain.

Some serious consequences appear to arise from the present results. In material composites involving glass fibers, properties strongly affected by interfacial conditions should be particularly sensitive to the selection of surface modifying agents. Adhesion at matrix/fiber interfaces is an obvious case in point, as are the mechanical properties of the system at high load; that is, in the region of non-linear response. The IGC method, and notably its ability to offer comparative indexes of acid/base activity, is useful as a guide to preferred surface

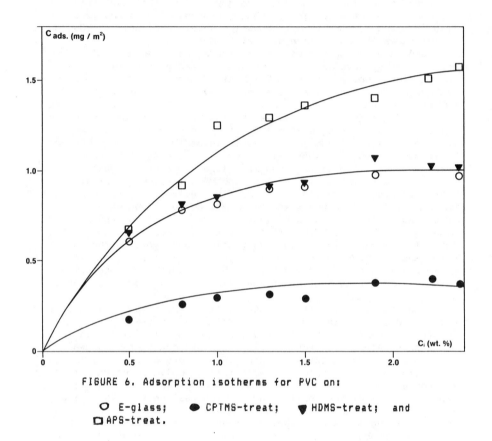

FIGURE 6. Adsorption isotherms for PVC on:

○ E-glass;   ● CPTMS-treat;   ▼ HDMS-treat;   and
□ APS-treat.

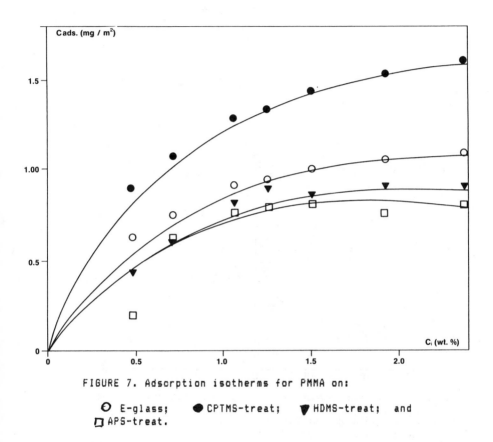

FIGURE 7. Adsorption isotherms for PMMA on:

⊘ E-glass; ● CPTMS-treat; ▼ HDMS-treat; and
□ APS-treat.

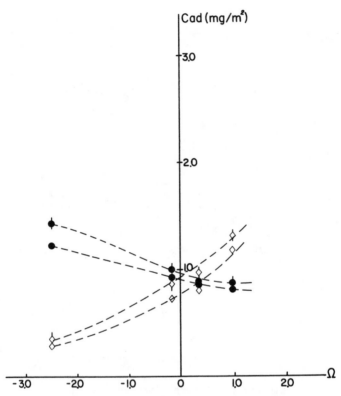

FIGURE 8. Polymer adsorption on glass fibers, as function of
acid/base concept:
PVC from 1% (◊) and 1.5% (◆) solution;
PMMA from 1% (●) and 1.5% (♦) solution.

treatments for reinforcing fibers. Thus, treatments leading to strongly basic surfaces may be preferentially selected for matrixes, such as PVC, which register as Lewis acids. However, basic matrixes may benefit from surface treatments rendering glass fibers acidic. Strong acid/base coupling, thereby designed into composite structures, should benefit property retention during use of the articles. Finally, the strong temperature dependence of interactions is again noted: the aging or property loss of polymer systems may be the result, at least in part, of temperature fluctuations, which necessitate composition adjustments at molecular or domain levels.

Acknowledgments
In part, this work was supported by grants from the Natural Sciences and Engineering Research Council, Canada. We are grateful to Owens-Corning Fiberglass, Granville, OH for its support and for the supply of screened, sized, glass fiber samples. Useful discussions with Dr. Sheldon P. Wesson, Textile Research Institute, Princeton, N.J. are particularly noted.

Literature Cited
1. Braun, J.M.; Guillet, J.E. Adv. Polym. Sci. 1976, 21, 108.
2. Gray, D.G. Prog. Polym. Sci. 1977, 5, 1.
3. Deshpande, D.D.; Patterson, D.; Schreiber, H.P.; Su, C.S. Macromolecules 1974, 7, 530.
4. DiPaola-Baranyi, G.; Richer, J.; Prest, W.M. Jr., Can. J. Chem 1985, 63, 223.
5. DiPaola-Baranyi, G.; Degre, P. Macromolecules 1981, 14, 1456.
6. Olabisi, O.; Robeson, L.M.; Shaw, M.T. Polymer-Polymer Miscibility; Academic Press: New York, 1979, chapter 3.
7. Boluk, Y.M.; Schreiber, H.P. Polym. Comp. 1986, 7, 295.
8. Schreiber, H.P. in Proc. XIII Internat. Conf. in Org. Coatings Sci.Tech., Athens, Greece, July 1987, p. 367.
9. Carre, A.; Gamet, D.; Schultz, J.; Schreiber, H.P. J. Macromol. Sci., Chem. 1986, A-23, 1.
10. Saint-Four, C.; Papirer, E. J. Colloid Interface Sci. 1983, 91, 69.
11. Chabert, B.; Chauchard, J.; Lachenal, G.; Philibert, T.; Soulier, J.P. Comptes Rendus Acad. Sci. 1982, 295, 987.
12. Drago, R.S.; Vogel, G.G.; Needham, T.E. J. Amer. Chem. Soc. 1971, 93, 6014.
13. Cuckor, P.M.; Prausnitz, J.M. J. Phys. Chem. 1972, 76, 598.
14. Fowkes, F.M.; Mostafa, M.A. IEC Prod. R&D. 1978, 17, 3.
15. Shafrin, E.G.; Zisman, W.A. J. Amer. Ceram. Soc. 1967, 50, 487.
16. Alexander, G.; Bradshaw, S.; Dodd, K.; Guthrie, J.T.; Mason, T. In Proc. XIII Internat. Conf. in Org. Coatings Sci. Tech. Athens, Greece, July, 1987, p. 133.
17. Fowkes, F.M. J. Adhesion Sci. Tech. 1987, 1, 7.

RECEIVED October 26, 1988

Chapter 18

Analysis of Solid Surface Modification

Eugène Papirer, Alain Vidal, and Henri Balard

Centre de Recherches sur la Physico-Chimie des Surfaces Solides, Centre National de la Recherche Scientifique, 24 avenue du Président Kennedy, 68200 Mulhouse, France

Inverse gas chromatography (IGC) is used for the determination of the surface energy characteristics of silicas before and after modification by heat treatment or by grafting onto their surface alkyl, poly(ethylene glycol) and alcohol chains. Because of its high sensitivity, IGC reveals the nature of the grafted molecules, which may then be confirmed by independent methods.

The addition of finely divided solids to rubber matrices is commonly practiced to increase the performance and service life of these materials. Indeed, without an active filler, a synthetic elastomer like Styrene Butadiene Rubber (SBR)would not be of much use. For instance, a tire made of pure vulcanized SBR would not last more than a few hundred miles. The introduction of coarse filler particles, such as milled quartz or clays, improves the situation so that the tire lasts thousands of miles. However, using active fillers like special grades of carbon black or silica has produced modern tires that operate satisfactorily for tens of thousands of miles.

The reinforcement of rubber by the presence of active fillers is a complex phenomenon that depends on the characteristics of the elastomer network and the properties of the fillers. The influential properties are the particle size, the morphology of particle aggregates, and the surface properties. The role of the geometrical characteristics of the filler is well understood, whereas the significance of the surface properties is more difficult to analyze. This situation stems essentially from the lack of adequate methods to analyze the surface of such small particles and from the fact that fillers differ from each other and need to be considered individually.

It is usually not necessary to change the surface activity of carbon blacks, whereas silicas demand special attention. For instance, it is necessary to treat silica before its use in SBR. Coupling agents like γ-mercapto propyl triethoxy silanes allow the formation of strong bonds between silica and the polymer. However, strong chemical bonds are not always desirable. This is typically the case for silica/silicone rubber mixes where strong and unavoidable links lead to a hardening of the mix, which becomes brittle and cannot be reworked. In this case, a surface deactivation treatment of the silica is essential.

The examples given above indicate the necessity of having a better understanding of the surface properties of divided solids that have received a surface treatment. The objective of this paper is to demonstrate how advantageous inverse gas chromatography (IGC) is in achieving this goal.

0097–6156/89/0391–0248$06.00/0
© 1989 American Chemical Society

Materials and Methods

Silicas produced by two processes have been investigated: three samples are representative of the hydropyrogenation process (Aerosil 130, Aerosil 200 and Aerosil 300 from Degussa and referred to as Silica A130, A200 and A300), two samples were prepared by a wet, precipitation process (Z 130 and Z 175 from Rhône Poulenc and referred to as P1 and P2). These silicas have surface areas of 130, 200, 300, 130 and 175 m^2/g respectively, and have a particle size too small to be used in a IGC experiment. Hence, they were agglomerated in an infrared die, crushed, and sieved (100 to 250 μm). Approximately 0.5 g of silica were introduced into stainless steel columns 50 cm long and 2.17 mm in diameter. Helium was used as the carrier gas at a flow rate of 20 ml/min. Before each measurement, the columns are conditioned at 150°C for 24 h. Symmetrical retention peaks were observed with alkanes. For other peaks, an integrator was used to determine the first order moment.

The silicas were modified by grafting alkyl chains, diols, or poly(ethylene glycols) (PEG). Since the hydroxyl groups of the silica are weakly acidic, the grafting reaction corresponds to an esterification:

$$Si - OH + ROH \longrightarrow Si - OR + H_2O$$

Since esterification is an equilibrium reaction, the treatment was performed using an excess of alcohol. The grafting of alcohols or diols was performed in an autoclave at 150°C. For PEG, special care was taken to avoid oxidation (outgassing of the silica/PEG mix and treatment under N_2 in a sealed tube at 150°C). In each case, the excess reagent was eliminated either by heat treatment under vacuum (volatile alcohols and diols), or by solvent extraction (THF) in a Soxhlet extractor.

Grafting ratios were calculated either from weight loss of grated silicas when heat treated in air at 750°C or from elemental analysis of the modified silica. The two methods give concordant results.

Results and Discussion

The surface chemistry of silica is, at first sight, relatively simple . Only two types of surface groups are possible: the hydroxyl or silanol groups and the oxygen double bridges or siloxane groups. However, free silanols (either isolated or geminal when two hydroxyls are located on the same silicon atom) and associated silanols (adjacent silanols bridged by H-bonding) have different chemical reactivities resulting in different contributions to the surface properties of the oxide.

London Component of the Surface Energy of Heated Treated Silicas. Surface energy is usually considered as the sum of two components: the London component (γ_S^L), steming from London forces, and the specific component (γ_S^{SP}), originating from all other types of forces (polar, H-bonding, metallic, etc). Two methods are commonly used for the measurement of surface energies: wettability and adsorption techniques.

The first method, wettability, can be evaluated from the contact angle of a drop of liquid deposited on the flat surface of the solid. This method hardly applies to powders like silicas because special care must be taken to control the surface porosity of a silica disk made from compressed silica particles. For a chromatographic silica, Kessaissia et al. (1) determined a γ_S^L value close to 100 mJ/m², whereas the polar component of the surface energy was found to be 46 mJ/m². Hence, the silica exhibits a large surface energy.

The second method of γ_S^L determination is based on the interpretation of adsorption isotherms of either the total isotherm (calculation of the spreading pressure) or the initial or linear part of the isotherm. IGC readily provides the necessary information (2).

An IGC method for the analysis of divided solids and fibers has been initiated by Gray et al. (3). It is illustrated here by the results obtained from the precipitated silica sample (P1). Injecting a series of n -alkanes at infinite dilution (at the limit of detection by the flame ionization detector) usually results in a linear variation of the logarithm of the net retention volumes (V_N) with the number of carbon atoms in the n-alkanes. This is illustrated in Figure 1 for measurements performed between 71 and 130°C. Thermodynamic considerations show that V_N and the standard free energy of adsorption of the alkanes are related by

$$\Delta G_A^o = - RT \, Ln \, V_N + B$$

where B is a constant, depending on the choice of reference states of the alkanes in the gaseous and adsorbed states. Thus, experimental observation allows the calculation of an incremental value corresponding to the ΔG_{CH_2} of adsorption of one - CH_2 - group. Further, when measuring ΔG_A^o as a function of temperature, ΔH_A values are calculated.

Assuming that ΔG_A^o and ΔH_A vary linearly with the number of carbon atoms, and taking into account the relation of Fowkes (4) for the calculation of the interaction energy through London forces Gray et al. (3) established the following equation :

$$\frac{\Delta G_{CH_2}}{N.a} = 2 \sqrt{\gamma_S^L \, \gamma_{CH_2}}$$

N.a transforms free energy units into surface energy units, N being Avogadro's number and "a" being the area of an adsorbed - CH_2 - group.γ_{CH_2} is the surface energy of a solid made only of -CH_2- groups; that is polyethylene (PE). Hence, all terms are either known (N, a, γ_{CH_2}) or measurable (ΔG_{CH_2}), except the quantity of interest: γ_S^1.

This method was first applied to follow the surface energy characteristics of silica samples prepared by heating, up 700°C, A200 and P2, that is silicas of different origins but comparable surface areas.

Gravimetric measurements of the weight losses during heat treatment indicated a smooth evolution of the weight: silica P2 lost much more water than silica A200. Nevertheless, the γ_S^1 measurements via IGC at 60°C indicated (Fig.2) a more complex variation with heat treatment. γ_S^1 increases dramatically when increasing the temperature up 500°C and then decreases. Both silicas follow similar trends, but significant differences show up between silicas A200 and P2. Surface silanol content measurements, made either by esterification with $^{14}CH_3OH$ or using alkyl aluminium derivatives, point to a progressive elimination of the hydroxyl groups. However silanol groups (approximately 1.5 group/nm^2) are still present despite the 700°C heat treatment. Therefore the variation of γ_S^1 cannot be justified only by the total concentration of surface hydroxyl groups.

In fact Maciel et al. (5) have shown using solid state NMR that variation of total silanol and geminal silanol contents are not at all connected. The fraction of geminal silanol groups changes during heat treatment in the same complex way as do γ_S^1 values. The change in γ_S^1 values of silica heated above 500°C is possibily related to reorganization ability of silica surface (6). Indeed, at temperatures above 500°C, sufficient thermal energy is provided to the silica network to allow relaxation of the highest strained siloxane bridges created by condensation of the silanol groups below 500°C.This relaxation is accompanied by variations in surface properties of silicas as demonstrated by IGC.Finally, it is seen that even though IGC is not able to reveal

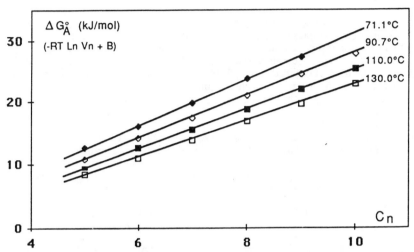

Figure 1. Variation of the net retention volume (V_N) of n-alkanes with number of carbon atoms, measured at different temperatures (column containing precipitated silica P1).

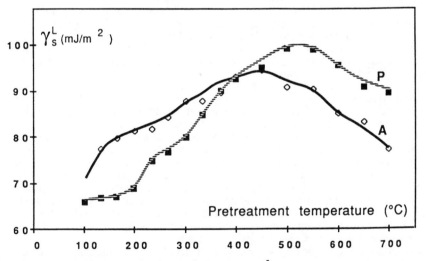

Figure 2. Variation of the London Component (γ_s^L) of precipitated (P) and fumed (A 200) silicas upon heat treatment.

the exact mechanisms of the chemical surface processes induced by thermal treatment, it appears as a most sensitive method to detect changes in the surface properties which are associated with these processes.

London Component of the Surface Energy of Silicas Having Alkyl Grafts.Some results pertaining to samples obtained by esterification of silica P1 are presented as an example in Figure 3.The data indicate that for these samples, the linear relationship between $\Delta G°$ and ΔH is accurate.consequently, the γ_s^L values may be determinated according to the method outlined earlier. Table I compares the γ_s^L of silicas before and after esterification either with short chains (methyl: C_1) or long chains (hexadecyl: C_{16}), A and P corresponding to the fumed(A130) and precipitated (P1) silicas, respectively.

Table I: London Component of the Surface Energy of Silica (P1 and A130)
            Before and After Methyl (C_1) and Hexadecyl (C_{16}) Esterification

	P	PC_1	PC_{16}	A	AC_1	AC_{16}	PE
Chains nm^2	-	10.5	2.2	-	3.4	1.5	-
γ_s^L (mJ/m^2)	98	87	47	75	70	38	35-40

The reaction with CH_3OH also allows the measurement of the silanol amounts. Silicas P and A have different contents. In fact, the content of P is so great that it exceeds the optimum surface coverage capacity (monolayer of - OCH_3 groups). Two hypothesis may be proposed to explain this result. Either the surface of the P silica is rugged, or non-condensed polysilicic acids chains are still present on silica P. Such pendant chains could react efficiently with methanol indicating an apparent excessive value of silanol surface coverage.

Grafting methyl chains onto the silicas decreased only slightly the γ_s^L values, a result accounted for by the small size of the - CH_3 group, which is unable to screen efficiently the silica surface. However, when attaching longer alkyl chains, γ_s^L is greatly decreased and approaches values close to that of polyethylene.

An interesting observation can be made when studying silica (A130) samples modified by grafting alkyl chains of increasing number of C atoms (Figure 4). A striking variation of γ_s^L is recorded with grafts having 7 or 11 carbon atoms. These results are beyond experimental errors, since the γ_s^L measurement is reproducible to within ± 0.5 mJ/m^2. In Figure 5, γ_s^L is plotted against the number of - CH_2 - groups per unit surface area (nm^2). The minima now correspond to approximately 15 CH_2/nm^2 and 23 CH_2/nm^2. Taking 0.06 nm^2 as the mean area of a - CH_2 - groups, it becomes obvious that minima are observed when the surface is covered by one and two monolayers of - CH_2 - groups, respectively.

Cross polarization, magic angle spinning solid state ^{13}C NMR measurements were performed (7) on the series of samples, examining more precisely the mobility of the end methyl group of the alkyl graft. NMR indicates that the mobility of this group is most restrained when either the monolayer or the second layer of -CH_2- groups are completed on the silica surface.

These results suggest that on silica A130, the grafted alkyl chains organize themselves so as to form a dense and regular array of -CH_2- groups having optimum interaction capacity both with the silioca surface and with neighbouring -CH_2-

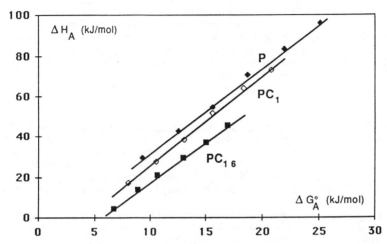

Figure 3. Relation between $\Delta H°$ and $\Delta G°$ of adsorption of n-alkanes on precipitated silicas (P1), methylated (PC_1) and hexadecylated (PC_{16}) silica samples.

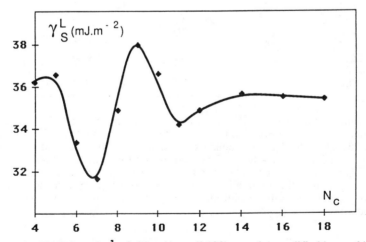

Figure 4. Variation of γ_S^L of silica (Aerosil 130) samples modified by grafting alkyl chains of increasing chain length (N_C).

groups. Clearly then , under these conditions, the interaction potential of the grafted chains with alkanes, used for IGC, are also reduced. Thus the observed minima in γ_S^l values are explainable.

London Component of the Surface Energy of Silicas Modified with Diols. These samples were prepared to compare alkyl grafts, formed during the reaction of silica with alcohols, and similar grafts having a hydroxyl group at their free end. Figure 6 agrees with the results exhibited in Figure 5; that is, γ_S^l passes through a minimum value when a monolayer of $-CH_2-$ groups is completed on the silica surface. The second minimum, approximately at $26-CH_2-/nm^2$, is less obvious than the previous one. This result is explained by the same considerations as those presented for silicas having alkyl grafts, considerations which also supported by NMR measurements ($\underline{7}$).

London Component of the Surface Energy of Silicas Having PEG Grafts. The dependence of the surface properties of grafted silica on the number of monomer units is also evidenced with silicas (A300) having poly(ethylene glycol) chains attached. Table II presents the γ_S^l values, measured by IGC at 60°C. The other quantities listed are: the molecular weight (Mw) of PEG graft ; τ, the grafting ratio (weight percent of PEG on silica); and n_{MU}, the number of monomer units per unit surface area (nm^2).

Table II: γ_S^l Values Measured at 60°C on Silica A 300 Modified by Grafting of PEG

Mw	τ (%)	γ_S^l (mJ/m^2)	n_{MU}
0	0	76	0
2,000	5	43	2.6
4,000	7	36	3.7
2,000	17	38	8.9
2,000	23	36	12.1
2,000	56	30	29.5
10,000	57	30	30.0
4,000	72	30	37.9
10,000	75	32	39.5

The value of γ_S^l drops significantly from 76 approximately to 30 to 38 mJ/m^2 when going from the untreated silica to a sample modified by 7 % PEG having a molecular weight of 4.000. This amount corresponds to 3.7 monomer units ($- CH_2 - CH_2 - O -$) nm^2; that is, a value sufficient to form a monolayer. A limit value of 30 mJ/m^2 is reached for higher surface coverage. Hence the pertinent factor, when considering grafting PEG onto silica is, not the molecular weight of the graft but rather, as previously outlined with the alkyl grafts, the surface coverage by the monomer units.

Specific Component of the Surface Energy of Silica having Alkyl Grafts. So far, the focus has been on measuring the γ_S^l values of silicas by IGC. The remainder part of this paper is devoted to the determination of the specific component of the surface energy. A simple method for the determination of the specific component of the surface energy, starting from IGC results, does not exist. However, several attemps have been made ($\underline{8,9,10}$) to evaluate, through IGC, specific interaction parameters of polar probes with polar surfaces.

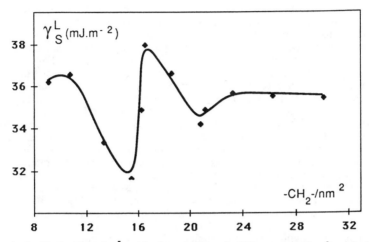

Figure 5. Variation of γ_S^{L} with the number of -CH$_2$- groups/nm^2 grafted on the surface of Aerosil 130 using alcohols as reactants.

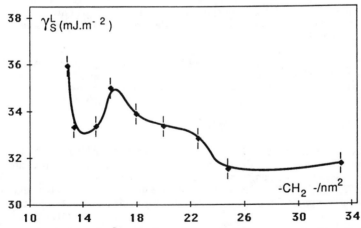

Figure 6. Variation of γ_S^{L} with the number of -CH$_2$- groups/nm^2 grafted on the surface of Aerosil 130 using diols as reactants.

For instance, a possible method to evaluate the contribution of specific interactions consists of the comparison of the chromatographic behaviour of two solutes having similar sizes (cyclohexane and benzene), yet different interaction capacities. Figure 7 illustrates this concept where the difference in ΔG_A of benzene and cyclohexane are plotted for three silica (P1) samples at various temperatures. The reference state of the adsorbed molecules is the same as that used by de Boer (11). As expected, the difference is greatest with the untreated silica. Yet, on PC$_{16}$, there persists a small possibility of specific interactions.

The previous method is essentially qualitative and does not allow prediction of the specific interaction potential of the silicas with other solutes. For a more quantitative approach, a semi-empiric method was developed to extract from the single chromatographic peak (or ΔG_A^0): the contribution of either London or specific interactions to the net retention volume V_N. The proposed method is illustrated by the following three figures corresponding respectivily to silica P1 and P1 samples having methyl and hexadecyl grafts. The first figure relates ΔG_A^0 to the vapor pressure of the injected solutes (Figure 8). This variable was chosen because it is pertinent thermodynamically. All n-alkane probes define a single straight line. By definition, the deviation from this line is taken as an estimation of the specific interaction parameter I_{sp}. A comparison of Figures 8 and 9 does not indicate any major differences, which confirms the fact that the methyl graft is too small in size to shield the silica surface. However, an examination of the results in Figure 10 shows significant differences, since the points corresponding to polar probes are close to the alkane line.

When comparing with the results obtained with hexadecylated silica A130 a major difference in behaviour is noted. For silica AC$_{16}$, all experimental points fit this alkane line. This result, reinforced by others (7,13) using techniques such as NMR and IR, is explained by assuming that the alkyl chains on A and P silicas are distributed differently. A regular array, which restricts the approach of the polar solutes to the solid's surface, is postulated for silica A. A patchwork type of organization, which allows polar parts of the surface access to the polar solutes, is postulated for silica P.

Specific Component of the Surface Energy of Silicas having PEG Grafts It is possible to take a step further for a more quantitative description of the specific interactions, a solid and a polar probe are able to exchange. It is based on the use of acid/base scales and an equation, which has been proposed earlier by Saint-Flour and Papirer (9)

$$I_{sp} = (AN) \, C + (DN) \, C',$$

where I_{sp} is the specific energy of interaction defined earlier, (AN) and (DN) are the acceptor and donor number of the probes injected in the GC, and C and C' are the capacities of the solid to exchange base or acid type of interactions. When applied for example to PEG grafted silica, this concept demonstrates the influence of the grafting ratio on the surface properties of grafted silica. Initially acidic, the silica acquires more base-like character (due to the ether links of PEG) as the grafting ratio is increased(12).

The results of IGC presented so far demonstrate its ability to determine and evidence minor changes in surfaces properties of solids submitted to various treatments. The last section of this paper will show the potential of IGC for the detection of unexpected molecular arrangements of the grafts on the silica surface.

Enthalpies of Interaction of Polar Probes with Silicas modified with Diols. Whereas the treatment of silicas with alcohols leads to fixation of alkyl grafts, their modification with diols results in the grafting of hydrocarbon chains still having a

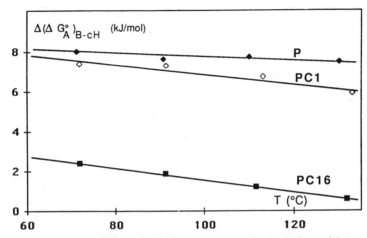

Figure 7. Comparison between the free energies of adsorption of benzene and cyclohexane on precipitated silice (P1), methylated (PC$_1$), and hexadecylated (PC$_{16}$) silica samples.

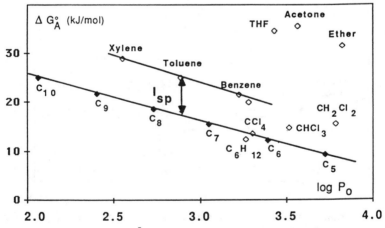

Figure 8. Variation of ΔG_A^o with logarithm of vapor pressure P$_O$ of probes (silica P1).

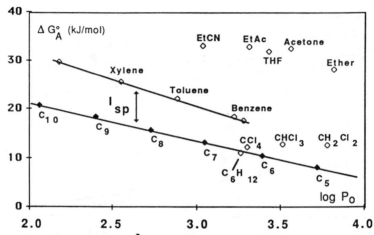

Figure 9. Variation of ΔG_A^o with logarithm of vapor pressure P_0 of probes (methylated silica P1).

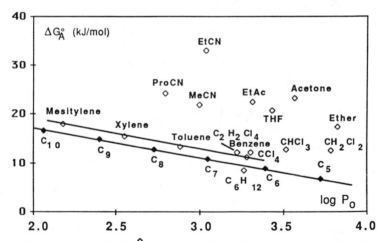

Figure 10. Variation of ΔG_A^o with logarithm of vapor pressure P_0 of probes (hexadecylated silica P1).

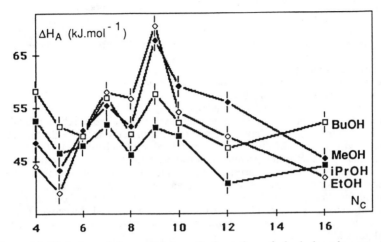

Figure 11. Variation of the enthalpies of adsorption of alcohol probes on silica A (A 130) samples modified by grafting diols of increasing chain length.

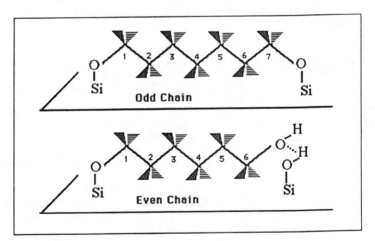

Figure 12. Schematic representation of the position of grafted diol chain on a flat silica surface.

terminal hydroxyl group. Hence, silica samples with different adsorption properties are possibily obtained and the study of their interaction capacity with hydrogene bonding probes should be most appropriate to evidence such differences.

Considering closely (Figure 11) the variation of the enthalpies of adsorption of alcohol probes on the surface of silica A130, which has been modified with diols of increasing chain length, a striking observation is made that seems to be related to the number of carbon atoms in the grafted diol. An explanation is proposed in Figure 12 which compares, in a schematic way, the configuration of grafted odd and even diol chains.

Several hypotheses are made:
- the surface of the silica is planar, on the molecular level;
- a diesterification is possible; and
- chains adopt a trans-trans configuration.

With these hypotheses, it is possible to understand the preferential diesterification reaction that occurs with diols having an odd number of carbons, since the terminal hydroxyl group of the odd diol is in a most favourable position. In fact NMR measurements (7) support the preferentiel diesterification reaction when using odd diols. The necessity of considering a flat surface is also demonstrated when comparing the results given by silicas A 130 and P1. Indeed, silica P has a more irregular surface, as can be shown by independent methods (13). Finally, the variations illustrated by Figure 11 are not observed with silica P1. Moreover for P1, NMR indicates essentially diesterification . All these facts are in favour of the proposed model.

According to this model, the variation of ΔH on mono and diesterified silica surfaces is accounted for by the greater capacity for H-bonding on the diesterified sample. On a mono-esterified silica, H-bonds already exists between the silanol and the terminal hydroxyl of the graft and does not facilitate the interaction with the alcohol probes.

Conclusion

IGC appears to be a useful and powerful method for the characterization of divided or fibrous solid surfaces. Because of its extreme sensitivity to small variations in the surface properties of the solid, IGC reveals interesting phenomena to be eventually confirmed by independent analytical methods.

This study shows that small and well defined molecules behave in a complex manner when chemically linked to the surface of a solid. Their behaviour is strongly dependent on the characteristics of the graft and the surface, and on geometrical factors like the fractality of the surface. Obviously, the grafting ratio, which determines the intensity of interactions that adjacent grafted molecules experience is also important.

It can be expected that the behaviour of a grafted macromolecule will be more complex to analyze. In addition to the aspects considered in this paper, one has to take into account the eventual modifications of the dynamics of the chains located above the polymer layer that is in direct contact with the solid.

Acknowledgment
The authors thank Miss Clara C. Pizaña for her kind and efficient assistance during the text editing.

Literature Cited

1. Kessaissia, Z.; Papirer, E.; Donnet, J.B. J. Colloid Interface Sci. 1981, 79 (1), 257-63.
2. Conder, J.R.; Young, C.L. In Physicochemical Measurements by Gas Chromatography, John Wiley: New York, 1979.

3. Dorris, G.M.; Gray, D.G. J. Colloid Interface Sci. 1980, 77 (2), 353-62.
4. Fowkes,F.M. J. Colloid Interface Sci. 1968, 28 ,493.
5. Sindorf, D.W.; Maciel, G.E. J. Am. Chem. Soc. 1983, 105,1487-1493.
6. Brinker, C.J.; Kirkpatrick, R.J.; Tallant, D.R.; Bunker, B.C.; Montez, B. J. Non-Cryst. Solids 1988, 99, 418-428.
7. Tuel, A.; Hommel, H.; Legrand, A.P.; Balard, H.; Sidqi, M.; Papirer, E. submitted to Chromatographia.
8. Schreiber, H.P.; Richard, C.; Wertheimer, M.R. In Physicochemical Aspects of Polymer Surfaces; Mittal, K.L., Ed.; Plenum Publ. Co.: New York; Vol 2, p 739.
9. Saint Flour, C.; Papirer, E. Ind. Engn. Chem. (Prod. Res. Dev.) 1982, 21 (4), 666-70.
10 Schultz, J.; Lavielle, L.;: This volume.
11. De Boer, H.J. The Dynamical Character of Adsorption, Oxford University Press: London, 1953.
12. Papirer, E.; Balard, H.; Rahmani, Y.; Legrand, A.P.; Facchini, L.;Hommel, H. Chromatographia 1987, 23 (9), 639-647.
13. Zaborski, M.; Vidal, A.; Papirer, E.; Morawski, J.C. submitted to Makromol. Chem.

RECEIVED September 29, 1988

ANALYTICAL APPLICATIONS

Chapter 19

Characterization of Siloxane Polymer Solvents by Family Regression of Gas Chromatographic Retentions of Aliphatic and Aromatic Probe-Solutes

R. J. Laub and O. S. Tyagi[1]

Department of Chemistry, San Diego State University, San Diego, CA 92182

The "family-plot" retention data of 5 solutes with a series of poly(methylphenylsiloxane) stationary phases have been examined in terms of the saturation vapor pressure p_A^o of the solute, the methyl/phenyl ratio of the solvent, and the temperature. Plots of $\ln V_g^o$ against $\ln p_A^o$ for a given solute over a range of temperature were found to be linear, as were the "isothermals", that is, the retention/vapor-pressure plots for an homologous series of solutes at a constant temperature. The family-plot slopes exhibited by the n-alkane probe-solutes were also found to be very sensitive to the aromatic content of the polymers. Thus, it appears that the "family" technique of GC data reduction can be a useful tool for characterizing the physicochemical properties of (polymer) stationary phases.

A conceptual difficulty arises in characterizing polymer stationary phases with gas-liquid chromatographic probe-solute specific retention volumes (1), namely, since it is a matter of experience that V_g^o remains finite, the mole fraction-based solute activity coefficient $^x\gamma_A^\infty$ must asymptotically approach zero as the molecular weight of the polymer stationary phase M_B becomes large:

$$V_g^o = 273 R/^x\gamma_A^\infty \, f_A^o \, M_B \tag{1}$$

where the subscripts A and B designate probe-solute and stationary solvent, respectively; and where f_A^o is the bulk-solute fugacity. Patterson, Tewari, Schreiber, and Guillet (2) and Covitz and King (3) circumvented this problem by employing weight-fraction based activity coefficients:

$$^w\gamma_A^\infty = (M_B/M_A) \, ^x\gamma_A^\infty \tag{2}$$

[1]Permanent address: Regional Research Laboratory (Council of Scientific and Industrial Research), Hyderabad 500 007, India

such that:

$$V_g^O = 273R/^W\gamma_A^\infty \; f_A^O \; M_A \tag{3}$$

where $^W\gamma_A^\infty$ for a given solute approaches a constant as M_B tends to infinity. However, and while Equation 3 is certainly a useful innovation, there is then incurred the drawback that, since values of $^W\gamma_A^\infty$ can range from zero to infinity depending upon the values of M_A and M_B, their interpretation, even in the instance of ideal solutions, let alone in those cases where there are subtle deviations from ideality, is rendered somewhat ambiguous. Much the same can also be said of other forms of activity coefficient (4-6) irrespective of the standard state chosen for the solute (7-11).

THEORY

An alternative method of data reduction was reported early in the history of gas chromatography by Hoare and Purnell (12-15; see refs. 16,17 for recent applications), who considered the dependence of the specific retention volume on the solute saturation vapor pressure p_A^O. Thus, taking the view [now recognized to be naive (18); see later], that the observed mole fraction-based solute activity coefficient can be decomposed into "athermal" and "thermal" components (19-22):

$$^X\gamma_A^\infty = \; ^X\gamma_a^\infty \; ^X\gamma_t^\infty \tag{4}$$

followed by substitution of this relation for $^X\gamma_A^\infty$ in Equation 1, produces the expression:

$$V_g^O = 273R/^X\gamma_a^\infty \; ^X\gamma_t^\infty \; p_A^O \; M_B \tag{5}$$

where, for convenience, the approximation has been made at this point that the solute fugacity f_A^O can be replaced by its vapor pressure p_A^O without serious error (see also later). Recall next the Clausius-Clapeyron relation in exponential form:

$$p_A^O = \exp(-\Delta H^V/RT) \exp(C) \tag{6}$$

where ΔH^V is the molar heat of vaporization of the solute, which is assumed to be independent of temperature; and where C is a constant of integration. Now, since the excess Gibbs free energy, enthalpy, and entropy of mixing are related to the activity coefficient by:

$$\ln \; ^X\gamma_A^\infty = (H^E/RT) - (S^E/RT) \tag{7}$$

we can write that:

$$^X\gamma_t^\infty = \exp(H^E/RT) \tag{8}$$

and that:

$$^X\gamma_a^\infty = \exp(-S^E/R) \tag{9}$$

Multiplying Equation 8 by 6 then produces the result:

$$^X\gamma_t^\infty \; p_A^O = \exp(C) \exp[(H^E - \Delta H^V)/RT] \tag{10}$$

In addition, the heats of solution ΔH^S and vaporization ΔH^V are related by:

$$\Delta H^S + \Delta H^V = H^E \tag{11}$$

Thus, when $H^E = 0$, $\Delta H^S = -\Delta H^V$. Moreover, in instances where $H^E \neq 0$, $\Delta H^S = H^E - \Delta H^V$. Substituting this result into Equation 10 thereby yields:

$$^X\gamma_t^\infty p_A^o = \exp(C) \exp(\Delta H^S/RT) \tag{12}$$

We now define a constant $\underline{a}$ such that:

$$a = \Delta H^S/\Delta H^V \tag{13}$$

where, for $H^E = 0$, $\underline{a} = -1$. [Note that Purnell (15) defined $\underline{a}$ as $\Delta H_e^S/\Delta H^V$, where ΔH_e^S is the molar heat of evaporation of the solute from solution. Evidently, $\Delta H_e^S = -\Delta H^S$ in all cases. Also, when $H^E = 0$, $\Delta H_e^S/\Delta H^V = -\Delta H^S/\Delta H^V = 1$. This results in the sign of $\underline{a}$ being reversed in ref 15.] We next substitute the product $[(a)(\Delta H^V)]$ for ΔH^S in Equation 12, and then multiply both sides by unity, chosen as $[\exp(aC) \exp(-aC)]$, followed by rearrangements:

$$^X\gamma_t^\infty p_A^o = \exp[C(1 + a)] \exp\{-a[(-\Delta H^V/RT) + C]\} \tag{14}$$

Comparing the second exponential in Equation 14 with that in 6 leads to the expression:

$$^X\gamma_t^\infty p_A^o = (p_A^o)^{-a} \exp[C(1 + a)] \tag{15}$$

Upon further substitution of this result into Equation 5, followed by taking logs,

$$\ln V_g^o = a \ln p_A^o + \ln[(273R/^X\gamma_a^\infty M_B)] - C(1 + a) \tag{16}$$

Equation 16 is conveniently abbreviated to the form:

$$\ln V_g^o = a \ln p_A^o + k \tag{17}$$

where $\underline{k}$ is defined by:

$$k = \ln[(273R/^X\gamma_a^\infty M_B)] - C(1 + a) \tag{18}$$

I. Ideal Solutions. In this case $^X\gamma_A^\infty = {}^X\gamma_a^\infty = {}^X\gamma_t^\infty = 1$ and, hence, $\Delta H^S = -\Delta H^V$. Plots of $\ln V_g^o$ against $\ln p_A^o$ will therefore have slopes $\underline{a}$ of -1, whence Equation 16 reduces to:

$$\ln V_g^o = -\ln p_A^o + \ln(273R/M_B) \tag{19}$$

II. Athermal Solutions. Here $^X\gamma_t^\infty = 1$ and, hence, $\underline{a} = -1$; however, $^X\gamma_a^\infty \neq 1$ and, thus, $^X\gamma_A^\infty \neq 1$. Plots of $\ln V_g^o$ against $\ln p_A^o$ will therefore still have slopes equal to -1. Also, Equation 16 reduces to the expression:

$$\ln V_g^o = -\ln p_A^o + \ln(273R/^X\gamma_a^\infty M_B) \tag{20}$$

Parenthetically, the intercepts of family plots for athermal solutions differ from Equation 19 only in $^x\gamma_a^\infty$, such that:

$$\frac{V_g^O(\text{ideal})}{V_g^O(\text{ather})} = \frac{273 \, R/p_A^O \, M_B}{273 \, R/^x\gamma_a^\infty \, p_A^O \, M_B} = {}^x\gamma_a^\infty \tag{21}$$

$^x\gamma_a^\infty$ is also often calculated from the relation:

$$\ln {}^x\gamma_a^\infty = \ln (r^{-1}) + (1 - r^{-1}) \tag{22}$$

where $\underline{r} = \overline{V}_B/\overline{V}_A$, the ratio of the solvent and solute molar volumes. (Thus, when $r = 1$, $^x\gamma_a^\infty = 1$.)

III. Thermal Solutions. In these instances, since $^x\gamma_t^\infty \neq 1$, $\underline{a}$ cannot be equal to -1 unless, as is highly unlikely, the product $^x\gamma_a^\infty \, {}^x\gamma_t^\infty$ were coincidentally to be unity. The slopes of plots of ln V_g^O against ln p_A^O thus will likely differ from -1, since they must in any event correspond to the ratio of ΔH^S to ΔH^V. In addition, the intercept will be given by $\ln[(273R/^x\gamma_a^\infty \, M_B)] - C(1 + a)$. Further, since $\underline{a} \approx -1$, the term, $C(1 + a)$, will approximate zero, and the intercept hence will be invariant in $\underline{a}$.

In all of the above, the weight-fraction based activity coefficient can be substituted for that derived with mole-fraction units by replacing M_B with M_A. Also, fugacity effects can be taken into account by substituting f_A^O for p_A^O, where:

$$f_A^O = p_A^O \exp[p_A^O(B_{AA} - \overline{V}_A)/RT] \tag{23}$$

and where B_{AA} is the bulk-solute second-interaction virial coefficient (23,24).

Since the capacity factor k' of a solute is related to V_g^O via several constants, plots of ln k' against ln p_A^O should also be effective in gauging solution ideality (16). Moreover, there is then presented the considerable advantage of obviating measurement of the column content of stationary phase (whether volume or mass), all the requisite data being available directly from stripchart tracings. The methodology thus has immediate practical appeal in the study of polymer solutions, for which determination of the stationary-phase weight, w_S, required in any event for the calculation of specific retention volumes, poses considerable difficulties (25-27).

However, it is not immediately clear what other advantages, if any, the mode of data analysis represented by Equation 16 might have over those founded upon $^x\gamma_A^\infty$ or $^w\gamma_A^\infty$. We have therefore assessed in this work the family-plot behavior of several solutes with the Ohio Valley (OV) series of poly(methylphenylsiloxane) stationary phases, for which retention data of high accuracy are available for a variety of solutes (28-33).

RESULTS AND DISCUSSION

Table I presents the slopes, $\underline{a}$, and intercepts, $\underline{k}$, of n-pentane, n-hexane, n-heptane, benzene, and toluene probe-solutes with the methylphenylsiloxane polymer solvents: OV-1 (0 mol % phenyl), OV-3 (10%), OV-7 (20%), OV-11 (35%), OV-17 (50%), OV-22 (65%) and OV-25 (75%) (28-33). Table II then

Table I. Slopes, a, and Intercepts, k, of Family Retention Plots (Cf. Equation 16) of Indicated Solutes with OV Stationary Phases at 30–80°C

Solvent	OV-1		OV-3		OV-7		OV-11		OV-17		OV-22		OV-25	
% Phenyl	0		10		20		35		50		65		75	
Solute	$-a$	k	$-a$	k	$-a$	k	$-a$	k	$-a$	k	$-a$	k	$-a$	k
n-Pentane	0.978	10.5	0.955	10.2	0.932	9.95	0.916	9.63	0.897	9.29	0.810	8.39	0.670	7.20
n-Hexane	0.967	10.3	0.966	10.2	0.944	9.93	0.931	9.64	0.920	9.36	0.849	8.63	0.761	7.90
n-Heptane	0.963	10.1	0.971	10.1	0.957	9.91	0.940	9.61	0.939	9.38	0.876	8.75	0.819	8.27
Benzene	0.948	10.3	0.963	10.4	0.976	10.6	0.988	10.7	1.01	10.9	0.987	10.6	0.995	10.7
Toluene	–	–	0.971	10.4	0.984	10.5	0.996	10.6	1.02	10.7	0.989	10.5	0.993	10.5

Table II. Slopes, $\underline{a'}$, and Intercepts, $\underline{k'}$, of Isothermal Plots (Cf. Equation 16) of n-Alkane Solutes with OV Stationary Phases at 30–80°C

Solvent	OV-1		OV-3		OV-7		OV-11		OV-17		OV-22		OV-25	
% Phenyl	0		10		20		35		50		65		75	
t/°C	$-a'$	k'	$-a'$	k'	$-a'$	k'	$-a'$	k'	$-a'$	k'	$-a'$	k'	$-a'$	k'
30	0.840	9.65	0.949	10.2	1.02	10.5	1.08	10.7	1.14	10.8	1.19	10.8	1.28	11.1
40	0.839	9.60	0.946	10.2	1.01	10.5	1.08	10.7	1.14	10.9	1.19	10.9	1.26	11.2
50	0.837	9.55	0.943	10.1	1.01	10.5	1.07	10.7	1.13	10.9	1.18	11.0	1.24	11.2
60	0.833	9.48	0.939	10.1	1.00	10.5	1.07	10.8	1.12	11.0	1.17	11.0	1.22	11.1
70	0.829	9.40	0.933	10.1	0.996	10.4	1.06	10.8	1.12	11.0	1.16	11.0	1.19	11.2
80	0.824	9.32	0.932	10.0	0.994	10.4	1.06	10.8	1.11	11.0	1.15	11.1	1.17	11.1

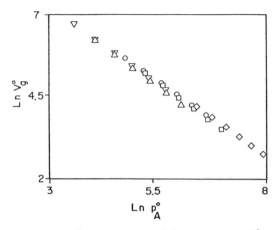

Figure 1. Family plots (cf. Equation 16) for n-pentane ($\diamond$), n-hexane ($\square$), n-heptane ($\triangle$), benzene ($\bigcirc$), and toluene ($\triangledown$) solutes with OV-1 stationary phase at 30-80°C.

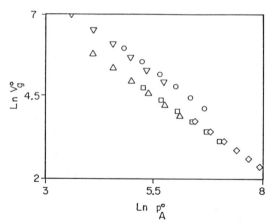

Figure 2. As in Figure 1; OV-11 stationary phase.

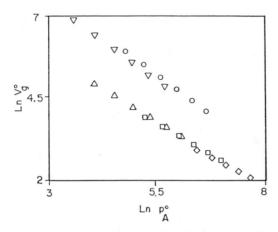

Figure 3. As in Figure 1; OV-25 stationary phase.

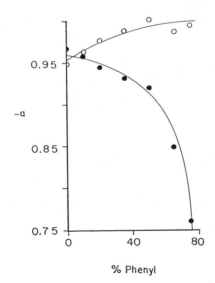

% Phenyl

Figure 4. Plots of family-regression slopes $\underline{a}$ against % phenyl content for n-hexane (filled circles) and benzene (open circles) with OV stationary phases at 30-80°C.

provides the slopes and intercepts of the isothermal plots, that is, $\ln V_g^0$ against $\ln p_A^0$, for the n-alkane probes at 30-80°C. The linear least-squares correlation coefficients $\underline{r}$ were in all instances greater than 0.9995. Typical family plots obtained with OV-1, OV-11, and OV-25 are shown in Figures 1-3, respectively.

From the standpoint of physicochemical measurements, family and isothermal plots are useful for the determination of vapor pressures (and, in addition, recalling Equations 13 and 16, heats of vaporization as well) from the retention data obtained from just a few chromatographic runs (33). Furthermore, the GC technique is ideally suited to instances in which the solutes are available only in minute quantities or are substantially impure, where each of these constraints ordinarily precludes bulk vapor-pressure measurements by conventional static procedures. For example, Heath and Tumlinson (34) employed log(retention) plots to determine the vapor pressures of trace acetate ingredients used in pheromone formulations. An important aspect of their work was that family correlations were obtained with a chiral-nematic stationary phase, cholesteryl p-chlorocinnamate.

Table I also shows that the $\underline{a}$ data for the aromatic hydrocarbons become more negative on passing from $\overline{OV}$-1 through OV-17, and fluctuate about -0.99 thereafter. That is, the retentions of benzene and toluene (weakly) reflect the aromatic character of the stationary phase up to a phenyl content of roughly 50%. In contrast, the values of $\underline{a}$ for the n-alkane probe-solutes are very sensitive to the aromatic content of the OV solvents, and increase sharply (become more positive) on passing from OV-1 to OV-25.

Plots of $-\underline{a}$ against phenyl content for n-hexane and benzene probe-solutes, provided in Figure 4, emphasize the dependency of the family-plot slopes of aliphatic and aromatic compounds (particularly those of the former) on the aromatic content of the OV solvents. The aromatic character of these stationary phases can therefore be gauged simply with the retentions of a few hydrocarbons. Presumably, with judicious choice of the probe-solute, other properties of polymer GC phases could be deduced from the slopes $\underline{a}$ of Equation 16 in a similar way. Family plots have also been shown recently to reflect discontinuities in retentions due to phase changes in mesomorphic polymers (16).

However, the derivation of Equation 16 failed to take into account the free-volume ("structural") contribution to $^x\gamma_t^\infty$ (35), nor is the effect negligible with some polymer phases (36). In addition, there remains considerable doubt even as to what constitutes a "family" set of data e.g., with polymers that are liquid-crystalline. The characterization of polymer stationary phases via family-plot regressions of "inverse" gas-chromatographic retentions therefore invites further and comprehensive study.

Acknowledgments

Support provided for this work in part by the Department of Energy Office of Basic Energy Sciences (analytical considerations) and by the National Science Foundation (physicochemical studies) is gratefully acknowledged.

Literature Cited

1. Laub, R. J.; Pecsok, R. L. Physicochemical Applications of Gas Chromatography; Wiley-Interscience: New York, 1978; pp 135-139.
2. Patterson, D.; Tewari, Y. B.; Schreiber, H. P.; and Guillet, J. E. Macromolecules 1971, 4, 356.
3. Covitz, F. H.; King, J. W. J. Polym. Sci., Part A-1 1972, 10, 689.
4. Huber, G. A.; Kovats, E. sz. Anal. Chem. 1973, 45, 1155.
5. Fritz, D. F.; Kovats, E. sz. Anal. Chem. 1973, 45, 1175.
6. Martire, D. E. Anal. Chem. 1974, 46, 626.
7. Ben-Naim, A. J. Chem. Educ. 1962, 39, 242.
8. Meyer, E. F. J. Chem. Educ. 1973, 50, 191.
9. Meyer, E. F. J. Chem. Educ. 1980, 57, 120.
10. Meyer, E. F. Am Lab. (Fairfield, Conn.) 1982, 14(10), 44.
11. Castells, R. C. J. Chromatogr. 1985, 350, 339.
12. Hoare, M. R.; Purnell, J. H. Research 1955, 8, S41.
13. Hoare, M. R.; Purnell, J. H. Trans. Faraday Soc. 1956, 52, 222.
14. Purnell, J. H. In Vapour Phase Chromatography; Desty, D. H., Ed.; Butterworths: London, 1957; pp 52-61.
15. Purnell, J. H. Gas Chromatography; Wiley: New York, 1962; Chapter 10.
16. Laub, R. J. Mol. Cryst. Liq. Cryst. 1988, 157, 369.
17. Laub, R. J.; Tyagi, O. S. Proc. Polym. Mater.: Sci. Eng. 1988, 58, 661.
18. Harbison, M. W. P.; Laub, R. J.; Martire, D. E.; Purnell, J. H.; Williams, P. S. J. Phys. Chem. 1979, 83, 1262.
19. Flory, P. J. J. Chem. Phys. 1941, 9, 660.
20. Flory, P. J. J. Chem. Phys. 1942, 10, 51.
21. Huggins, M. L. J. Phys. Chem. 1942, 46, 151.
22. Huggins, M. L. J. Am Chem. Soc. 1942, 64, 1712.
23. Laub, R. J. Anal. Chem. 1984, 56, 2110.
24. Laub, R. J. Anal. Chem. 1984, 56, 2115.
25. Laub, R. J.; Purnell, J. H.; Williams, P. S.; Harbison, M. W. P.; Martire, D. E. J. Chromatogr. 1978, 155, 233.
26. Laub, R. J.; Pecsok, R. L. Physicochemical Applications of Gas Chromatography; Wiley-Interscience: New York, 1978; pp 34-37.
27. Ashworth, A. J.; Chien, C.-F.; Furio, D. L.; Hooker, D. M.; Kopecni, M. M.; Laub, R. J.; Price, G. J. Macromolecules 1984, 17, 1090.
28. Chien, C.-F.; Kopecni, M. M.; Laub, R. J. J. High Resolut. Chromatogr. Chromatogr. Commun. 1981, 4, 539.
29. Chien, C.-F.; Furio, D. L.; Kopecni, M. M.; Laub, R. J. J. High Resolut. Chromatogr. Chromatogr. Commun. 1983, 6, 577.
30. Chien, C.-F.; Furio, D. L.; Kopecni, M. M.; Laub, R. J. J. High Resolut. Chromatogr. Chromatogr. Commun. 1983, 6, 669.
31. Chien, C.-F.; Kopecni, M. M.; Laub, R. J. J. Chromatogr. Sci. 1984, 22, 1.
32. Laub, R. J. J. High Resolut. Chromatogr. Chromatogr. Commun. 1987, 10, 565.
33. Laub, R. J.; Purnell, J. H. J. High Resolut. Chromatogr. Chromatogr. Commun. 1988, 11, in press.
34. Heath, R. R.; Tumlinson, J. H. J. Chem. Ecol. 1986, 12, 2081.
35. Janini, G. M.; Martire, D. E. J. Chem. Soc., Faraday Trans. 2 1974, 70, 837.
36. Flory, P. J.; Hocher, H. Trans. Faraday Soc. 1971, 67, 2258.

RECEIVED December 13, 1988

Chapter 20

Analyte Competition on Polyimide Adsorbents Studied by Deuterated Tracer Pulse Chromatography

James H. Raymer, Stephen D. Cooper, and Edo D. Pellizzari

Research Triangle Institute, 3040 Cornwallis Road, Research Triangle Park, NC 27709

Deuterated tracer pulse chromatography (TPC) was used to characterize the retention behavior of Tenax-GC and four polyimide-based sorbent materials. Deuterated n-hexane, ethanol, 2-butanone, nitromethane, and benzene were used as compounds to probe five types of chemical interactions of the compounds with the polymers. Retention properties were investigated with dry and humidified helium carriers both with and without the incorporation of non-deuterated test compounds. Analyte competition was shown to occur on all of the sorbents. Humidity affected the retention of the probe compounds on the polyimides to a much greater extent than on Tenax-GC. The technique was shown to elucidate subtle differences in sorbent behavior.

The identification and quantification of organic compounds in ambient air are problems that are complicated by the wide range of molecular weights and polarities of these compounds, and by the trace levels at which these compounds are present. One of the most useful methods to overcome the problem of low analyte (target compound) concentration is using a chromatographic pre-concentration technique with an adsorbent such as Tenax-GC [porous poly(2,6-diphenyl-p-phenylene oxide)] (1,2). In such an analysis, an air stream is drawn through a cartridge packed with the adsorbent material, and the organic compounds are selectively retained. The trapped compounds are subsequently thermally desorbed and cryogenically focused onto the head of a gas chromatographic column for analysis. Recently, several groups have reported the use of supercritical fluids for extraction of adsorbents or environmental solids (3-7); some of these methods (6,7) describe the direct introduction and focusing of the extracted compounds into a gas chromatographic column. Although a further discussion of procedures that use supercritical fluids is beyond the scope of this paper, the investigation of such methods is indicative of the importance of adsorbent-based analytical methods for environmental analysis.

0097–6156/89/0391–0274$06.00/0

One limitation of adsorbent-based preconcentration, which manifests itself during the sample collection step, is the poor retention of certain compounds on the adsorbent itself. For example, Tenax-GC retains nonpolar compounds more effectively than polar compounds, such as methanol or vinyl chloride (2), making the quantification of such poorly retained, polar materials difficult. Adsorbents that have higher affinities for polar compounds can help overcome this limitation.

The development of more polar, polyimide-based adsorbents was the goal of an earlier research project at Research Triangle Institute (8). Of the many polymers synthesized, the four depicted in Figure 1 were shown to provide good retention of polar compounds, good thermal stability (necessary for use with thermal desorption), and low background. Low background is used here to mean that few chromatographic peaks were detected during GC analysis of the compounds thermally desorbed from clean adsorbent. The goal of the current work was to evaluate the suitability of these polymers for use in air sampling, and to compare the results with those for Tenax-GC, one of the standard adsorbents.

In actual field operations, the air to be sampled can contain widely varying amounts of water and organic compounds. As air containing some constant level of organic compound is drawn through the adsorbent cartridge, this compound begins to accumulate and migrate through the adsorbent bed as in frontal chromatography. The volume of sampled air necessary to cause the migration of the front out of the end of the adsorbent bed is called the breakthrough volume (BV). It is extremely important that the volume of air sampled not exceed the breakthrough volume of the target compounds because the calculated levels in the air will be inaccurate. Breakthrough volumes can be affected by, among other things, the mass of a particular analyte and the presence of other analytes. High concentrations of analyte in the incoming air stream can cause the retention of the analyte to decrease from that observed at lower concentrations. Humidity is also expected to greatly affect retention when dealing with a polymeric material having a relatively high affinity for water. Therefore, the dependence of BV on humidity and the presence of large quantities of certain compounds, which might or might not be of interest, should be studied carefully before a given adsorbent can be used reliably in the field. Relevant considerations in the present study include the effects of humidity on the retention of analytes and competition among the analytes for sites on the adsorbent surface. It was desirable to gain insight into the mechanisms of adsorption, and to discern whether the adsorption process is related to a particular class of chemicals.

Two concerns dictated the choice of test analytes and experimental methods. The first concern was to choose compounds that would probe four types of molecular interactions that might occur during adsorption. In addition to water (to probe strong hydrogen bonding), the probe compounds used were ethanol (electron donor properties and weaker hydrogen bonding), 2-butanone and nitromethane (electron donor properties and no active hydrogen bonding), and benzene (pi-pi or induced dipoles forces and no active hydrogen bonding). n-Hexane was used as representative of aliphatic hydrocarbons, where London dispersion forces predominate in the

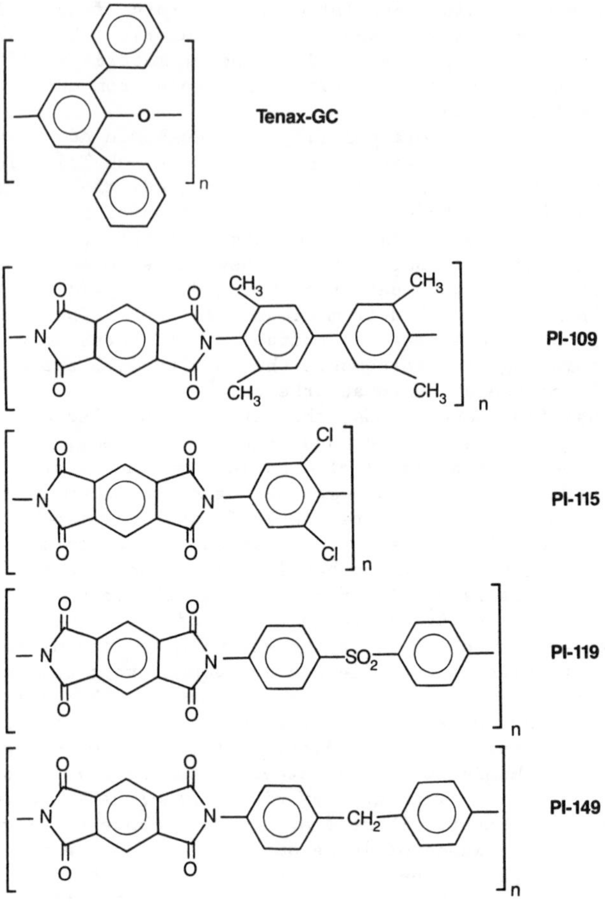

Figure 1.   Structures of Tenax-GC and polyimide sorbents examined in this study.

adsorption process, because such compounds are often present as
high-level background components in air. The second concern was to
accurately mimic field sampling conditions where the adsorbent is
initially "clean" and accumulates increasing amounts of material
with time, thus, possibly giving rise to variable competition
effects. A deuterated tracer pulse technique with mass
spectrometric detection, which is a modification of the method of
Parcher (9,10), was used in this study. The polymer was packed into
a gas chromatographic column, and thus defined this as an inverse
gas chromatography (IGC) experiment.

The use of Tracer Pulse Chromatography (TPC) is depicted in
Figure 2 for a single deuterated compound (probe) and its non-
deuterated analogue added as a front, that is, added at a known
level to the carrier gas. Figure 2A shows the case for duplicate
injections (pulses) of the deuterated probe with pure helium
carrier. The resulting two retention times are the same and serve
as the controls. Figure 2B shows the case where the front was
introduced (second arrow) slightly after the injection of the
deuterated compound. For purposes of illustration, the front and
the pulse are the same chemical compound. In Figure 2B, the time t_1
is the same as t_1 of Figure 2A, indicating the reproducible
migration of the compound with pure carrier. Notice that the second
injection of deuterated compound at t_2 results in the elution of the
component at t_3 which is less than t_1. That is, t_3-t_2 is less than
t_1 indicating a reduced retention. The dotted peak of Figure 2B
indicates the elution time if the front were absent as in Figure 2A.
The difference in elution times, Δt, represents the effect of
additional mass of the compound in question. More information is
obtained when the front and the probe are not the same compound.
Changes in retention in this case most likely represent competition
of the probe molecule and the front compound for similar sites on
the adsorbent surface. As changes in retention are observed for the
different probe compounds as various components of the front
accumulate on the column, information can be gleaned about
competition for adsorbent sites and, indirectly, about the sites
themselves.   A matrix of experiments was performed for each
polymer, as described in the next section. The resulting
information provides the basis for defining the actual sampling
conditions.

Materials and Methods

A diagram of the apparatus used to study the polymers is shown in
Figure 3. The polymer (40 to 60 mesh) was packed into a 2 mm i.d. x
80 cm glass gas chromatographic column and held at a constant
temperature in a Varian 3700 Gas Chromatograph (Varian Associates,
Walnut Creek, CA). Tenax-GC was purchased from Alltech Associates,
Applied Science Labs. A temperature was chosen that would allow
elution of all probe molecules within 30 minutes. The use of
different temperatures was assumed not to affect the adsorbent
properties. Based on our earlier study, these polyimide materials
have high glass transition temperatures (in excess of 280°C) and so
should be in approximately the same physical state at 135°C as at
ambient temperature. Detection was accomplished using an LKB 2091
mass spectrometer (LKB, Bromma, Sweden) operated in the multiple ion

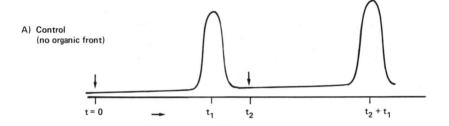

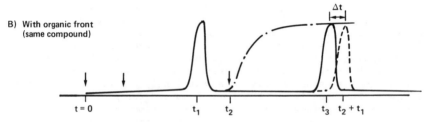

Figure 2.   Schematic Description of Dynamic Tracer Pulse
Chromatography. Arrows at t=0 and t_2 indicate
injections of deuterated "pulse" compounds. The
additional arrow in B indicates the initiation of the
organic front.

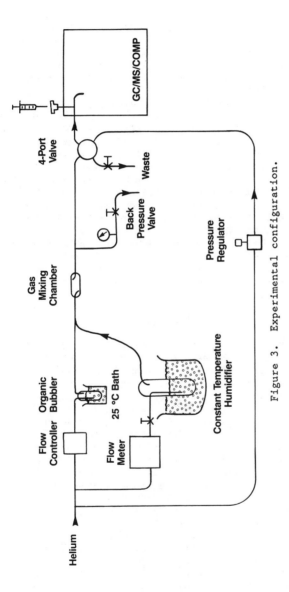

Figure 3. Experimental configuration.

detection mode. The carrier gas was helium and was used untreated
for control experiments, humidified using a constant temperature
humidifier to study the effects of moisture on retention, or passed
through a bubbler to introduce a front of the non-deuterated organic
probe compounds (approximately 100 nmoles/min) in the absence and
presence of 90% relative humidity. Discrete pulses of the
corresponding deuterated chemicals (approximately 5 μg each) were
injected into the carrier stream at 0, 10, 20, 30, and 40 minutes.
The frontal stream was begun at 10 minutes. Frontal streams
containing a 10-fold increase in the concentration of one compound
in the mixture were also used to isolate the effects of a particular
compound on the retentions of the probe molecules. Each compound
was elevated in turn. Frontal streams containing 10 and 40
nmoles/min of each compound were also used to ensure that changes in
retention measured with the high frontal levels were not
manifestations of adsorbent capacity. Changes in retention at lower
levels that are not seen at the higher level could also provide
clues to competitions that might otherwise be masked. The mass of
each pulse compound was chosen based on a plot of retention volume
versus mass of each compound injected into a pure carrier gas stream
as illustrated in Figure 4. A mass corresponding to the point on
the curve where retention volume begins to decrease rapidly with
increasing mass (arrow in Figure 4), should most readily show
changes in retention because of competitions. The percent change in
BV of the deuterated chemical was determined relative to an unspiked
frontal stream, that is, pure helium and no humidity. This
technique was effective in demonstrating relative changes in BV of ±
5%.

Results and Discussion

When the outlined matrix of experiments was performed for each
polymer, massive quantities of data were produced. For brevity and
clarity, the behavior of 2-butanone on Tenax-GC and PI-119 is
detailed, and the behavior of the other compounds on these two
polymers is summarized. The results for the other polymers are then
summarized. The results provided information relevant to the
questions posed at the beginning of this work: (a) does humidity
alter the BV of analytes on polyimides and Tenax-GC; (b) is BV for
an individual analyte altered by competition among analytes in a
mixture for the adsorption sites; and (c) is there a variety of
adsorption mechanisms and are these chemical-class related?
     Figures 5 and 6 show data for 2-butanone on Tenax-GC and PI-
119, respectively. In the figures, "% Deviation" refers to the
change in BV as compared with the control (pure, dry helium).
Therefore, a deviation of -10% means that the BV was decreased by
10%, relative to the corresponding control pulse, as a result of the
change in the system for that particular experiment. This method of
reporting the data elucidates the time-dependence of the retentions
of the probe compounds as compounds present in the front accumulate
on the adsorbent thus helping to model the environmental sampling
process. As an example, consider three compounds that, in any given
pulse, elute in the order A, B, and C. The fronts would break
through in the same order. For purposes of discussion, the
retention of compound B will be followed. Assume that an

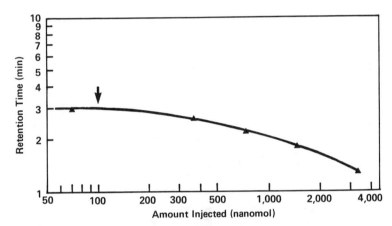

Figure 4. Retention time as a function of mass for n-hexane on PI-119. Arrow indicates approximate level of compound chosen for use in the pulse.

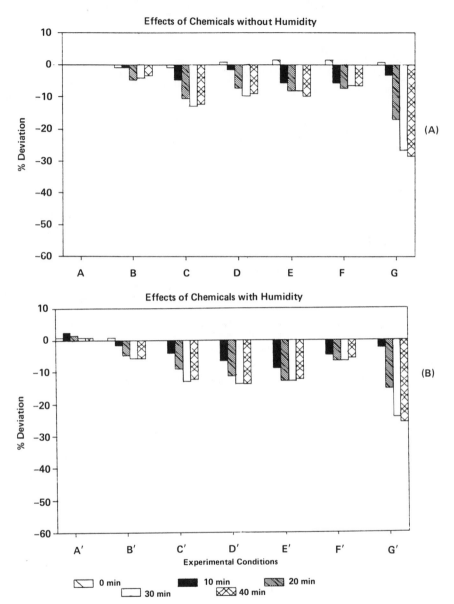

Figure 5.   Retention data for 2-butanone on Tenax-GC without (A)
             and with (B) humidity. Definition of letters on x-
             axis: A- no organic fronts; B- equimolar fronts at 80
             to 100 nanomoles each/min; C- elevated ethanol; D-
             elevated nitromethane; E- elevated n-hexane; F-
             elevated 2-butanone; G- elevated benzene.

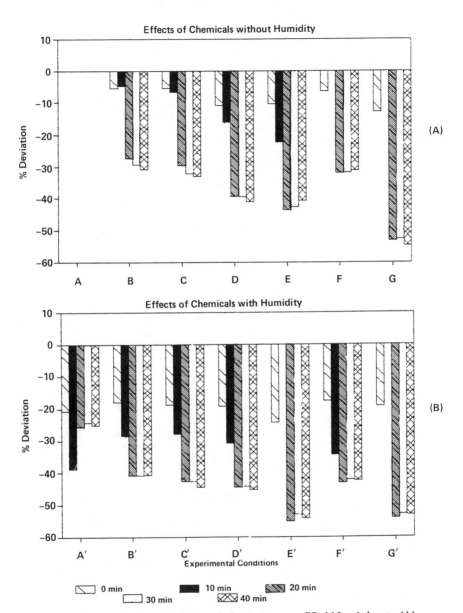

Figure 6.   Retention data for 2-butanone on PI-119 without (A) and with (B) humidity. Definition of letters on x-axis: A and B as in Figure 5. C- elevated n-hexane; D- elevated benzene; E- elevated ethanol; F- elevated nitromethane; G- elevated 2-butanone.

increasingly negative % deviation, as determined by comparing the
retention to the corresponding pulse in the absence of frontal
compounds (control), is observed for B until C breaks through the
adsorbent bed. Subsequent pulses for B show no further decreases in
retention. This suggests that compound C has a higher affinity for
the adsorbent than B and thus competes with B for surface sites. If
there were no overlap in sites of interaction for B and C, the
retention of B should be unaffected by C. Any competition of A with
B can only be elucidated by raising the level of A in the front and
repeating the experiment. Larger negative deviations for B in this
case would suggest competitions with A also. In reality, B is more
likely to displace A from the adsorbent because B has a higher
affinity for the surface as reflected by its longer retention.
Elution orders of the probe molecules were different on each
adsorbent.

Tenax-GC. For Tenax-GC at 85°C and 10 mL/min carrier flow, elution
order was: ethanol (1.9 minutes), nitromethane (7.6 minutes),
n-hexane (12.1 minutes), 2-butanone (13.9 minutes), and benzene
(17.7 minutes). Referring to Figure 5A, only elevated levels of
ethanol and benzene affect the retention of 2-butanone more than
10%. The observation that ethanol seems to compete to a small
extent with 2-butanone is reasonable because they both are electron
donors. The magnitude of this difference suggests that there is
only a small overlap in preferred adsorption sites. However, notice
that benzene competes strongly with 2-butanone, which suggests that
ethanol and benzene interact with different portions of the sorbent
molecule based on the magnitudes of the competitions with 2-
butanone. 2-Butanone tends to interact more with sites preferred by
benzene than with sites preferred by ethanol. An increased level of
2-butanone in the carrier stream (F in Figure 5A) affects the
retention of 2-butanone to a lesser extent than either ethanol or
benzene indicating that these interactions are chemical-class
related and not simply a result of the increased mass of 2-butanone.
    Figure 5B indicates that the addition of water vapor to the
carrier does not change the retention of 2-butanone to a large
extent relative to the dry carrier. This is consistent with the
known low affinity of water for Tenax-GC (2). Note that with
applied humidity, the affect on the retention of 2-butanone
associated with ethanol is no longer observed, and the effect of
benzene on the retention of 2-butanone is somewhat less than when no
humidity is present. This is consistent with the existence of a
small overlap in sorptive sites for ethanol and benzene, and with
the fact that water competes more effectively than either of these
two compounds. The overall results suggest that, when sampling for
2-butanone, high levels of a compound that experiences pi-pi
interactions can reduce the BV for 2-butanone, and that water vapor
has a relatively small effect.
    The retentions of n-hexane and benzene on Tenax-GC were also
affected (reduced BV) by the presence of organic compounds in the
carrier gas stream. n-Hexane appeared to be affected by increased
levels of every component and was more affected by components with
longer retentions; that is, compounds with more affinity for the
adsorbent. This result suggests that n-hexane undergoes a
relatively non-specific association with the surface. The mass of

adsorbed compound seems more important than the type of compound.
Humidity did not affect the retention of n-hexane and benzene to a
significant degree. The behavior of benzene was essentially the
same as that of n-hexane, only to a lesser extent; that is, changes
in retention were not as large as for n-hexane. It is logical that
benzene would be less affected than n-hexane because of benzene's
greater similarity to the structure of Tenax-GC (Figure 1).

PI-119. The elution order of the probes on PI-119 at 135°C and 10
mL/min carrier flow was: n-hexane (2.8 minutes), benzene (9.9
minutes), ethanol (10.8 minutes), nitromethane (16.7 minutes), and
2-butanone (13.3 minutes). Data for the retention of 2-butanone on
PI-119 are shown in Figure 6. The presence of the organic compounds
in the front greatly reduces the retention of 2-butanone, and this
effect grows as more material accumulates on the adsorbent.
Elevated levels of benzene (D in Figure 6A), ethanol (E), and 2-
butanone (G) reduce the retention volumes further, suggesting
compound-specific interactions. Figure 6B shows how humidity alone
(A) reduces the retention of 2-butanone; the addition of organics in
the front reduces retention more. Only elevated levels of ethanol
(E) and 2-butanone (G) cause the retention of 2-butanone to decrease
further (another 10%). The competition with benzene (D, Figure 6A)
has been masked by humidity (D, Figure 6B). Through the comparison
of Figures 5 and 6, it is evident that water vapor is an important
consideration in a sampling strategy utilizing a polar adsorbent.
    The results for the other probes are not as dramatic. In the
absence of humidity, the retention of ethanol was reduced
approximately 10% more than the reduction observed for low level
organics when benzene, nitromethane, and 2-butanone were elevated.
The presence of water vapor masked these differences and resulted in
retention volumes 30 to 40% lower relative to the control. The
retention of n-hexane seemed more affected by the total amount of
organics adsorbed, as was the case for Tenax-GC, than by increased
levels of individual organic compounds. Water vapor only affected
the retention of n-hexane approximately -10%. The behavior of
n-hexane with and without water vapor suggests an interaction of
n-hexane with the surface of the polyimide unhindered by the
adsorption of water. Benzene showed a decrease (approximately 10%
relative to organics alone) when benzene and 2-butanone were
elevated. The addition of humidity caused a slight reduction
(approximately 5%) in the retention of benzene and the addition of
the organics caused further (5 to 10%) reductions. In this case,
ethanol caused a reduction in benzene retention, whereas none was
seen in the absence of humidity. This suggests that water modified
the adsorbent surface such that ethanol then competed with benzene.
Although the retention of nitromethane was reduced approximately 10%
upon the introduction of water vapor, no dramatic effects on the
retention of nitromethane were induced by any of the other probe
molecules.

PI-109. The elution order of the probe molecules on PI-109 at 135°C
and 10 mL/min carrier flow was: n-hexane (4.8 minutes), ethanol
(9.9 minutes), benzene (15.7 minutes), nitromethane (17.6 minutes),
and 2-butanone (26.8 minutes). For n-hexane, humidity caused a 10%
reduction in retention volume and none of the compounds, when

elevated, indicated any compound-dependent change. In the absence
of humidity, the retention of ethanol was decreased approximately
10%, relative to the fronts alone, when elevated levels of ethanol,
benzene, nitromethane, or 2-butanone were in the front. The
addition of water vapor caused a generalized decrease in the BV of
ethanol between 20% and 30%. For benzene in the absence of
humidity, small decreases in retention volume were seen as a result
of elevated levels of the individual probe molecules, with the most
pronounced decrease associated with 2-butanone. Upon the
introduction of humidity, there was a generalized decrease in
retention with no particular probe causing a change larger than any
of the others. In this experiment, the retention volume of benzene
was smaller (large negative % deviation) when only water vapor was
introduced into the system than was seen when humidity plus organics
was introduced. A possible explanation is that the adsorbent
surface is modified substantially by the presence of water vapor,
and that the compounds from the front dissolve in the water at the
surface and interact with the benzene as yet another modified phase.
It is also possible that the compounds in the front displace water
from some of the sites and mask the effect of only water.

Deviations of 10 to 15% were observed in the retention of
nitromethane when elevated levels of nitromethane and 2-butanone
were introduced into the fronts with dry carrier. Similar effects
were observed for the retention of 2-butanone with elevated levels
of ethanol and nitromethane. The introduction of water vapor caused
a 10 to 15% decrease in the retention volumes of nitromethane and 2-
butanone for all organic front configurations. The greatest effects
on 2-butanone were induced by ethanol and 2-butanone, whereas there
were no clear compound-induced changes in the retention of
nitromethane. An initially large decrease in retention for
nitromethane and 2-butanone was measured when humidity alone was
added to the carrier; as was the case with benzene, this decrease
was less when the organic fronts were introduced.

PI-149. The elution order of the probes on PI-149 at 120°C and 10
mL/min carrier flow was: n-hexane (0.7 minutes), ethanol (7.7
minutes), benzene (13.1 minutes), 2-butanone (19.4 minutes), and
nitromethane (29.6 minutes). The investigation of retentions on PI-
149 indicated, as is the case for all polyimides, that water vapor
causes a reduction in retention volumes relative to the no humidity
situation. The only evident changes in retention, as a result of
the elevation of one compound in the front, was observed with
benzene after the elevation of benzene, 2-butanone, or nitromethane.

PI-115. The elution order on PI-115 at 90°C and 10 mL/min carrier
flow was: n-hexane (0.9 minutes), benzene (5.1 minutes), ethanol
(16.5 minutes), 2-butanone (10.7 minutes), and nitromethane (30.8
minutes). PI-115 showed large changes in its behavior with the
introduction of humidity. The behavior of benzene is interesting in
that the retention increased (positive % deviation) from the pulses
at 0 and 10 minutes, then decreased. This behavior is reproducible
because the same trend was observed in each experiment. Benzene,
ethanol, and 2-butanone caused the largest negative % deviation
values with 2-butanone, causing a maximum negative deviation at 40%
for the pulse at 40 minutes. This increasing then decreasing

retention trend was observed for humidity alone and in conjunction with the organic fronts, which suggests that benzene can displace water, and is thus retained longer. When a polar, longer retained compound migrates through the adsorbent bed, it displaces water and benzene, resulting in the reduced retention volume from subsequent pulses. The fact that this was not observed with PI-149 indicates the power of this technique to elucidate subtle aspects of the polymeric materials. The BVs for ethanol were decreased 50 to 60% and the largest effect was seen as a result of an increased ethanol level. For 2-butanone, the retention volumes decreased until all values were decreased by the 40 minute pulse. Benzene, ethanol, and 2-butanone caused the largest decreases in retention (-45% to -50% relative to the no-humidity case). Elevated levels of ethanol, 2-butanone, and nitromethane caused 35% to 40% reductions in the retention volume of nitromethane.

Conclusions

The dynamic tracer pulse technique used in this work facilitates the study of BV and how BV might be altered because of high levels of organic compounds or humidity. Based on competitions of the various, selected probes, cursory information about surface sites can be obtained for a prospective adsorbent; such information is especially useful for multifunctional polymers. This technique can also permit the fine tuning of the environmental collection strategy through the examination of retentions on mixed adsorbents or multiple adsorbent beds. The use of dynamic TPC with the polyimides demonstrated that these materials are more sensitive, in terms of BV of the tested probe molecules, to the effects of humidity than is Tenax-GC.

Acknowledgment
This project was supported by Grant R01 OH02108 from the National Institute for Occupational Safety and Health of the Centers for Disease Control.

Literature Cited

1.   Pellizzari, E. D.; Bunch, J. E.; Berkely, R. E.; McRae, J.
     Anal. Lett. 1976, 9, 45.

2.   Krost, K. J.; Pellizzari, E. D.; Walburn, S. G.; Hubbard, S. A.
     Anal. Chem. 1982, 54, 810.

3.   Raymer, J. H.; Pellizzari, E. D. Anal. Chem. 1987, 59, 1043.

4.   Raymer, J. H.; Pellizzari, E. D.; Cooper, S. D. Anal. Chem.
     1987, 59, 2069.

5.   Wright, B. W.; Wright, C. W.; Gale, R. W.; Smith, R. D. Anal.
     Chem. 1987, 59, 38.

6.   Wright, B. W.; Frye, S. R.; McMinn, D. G.; Smith, R. D. Anal.
     Chem. 1987, 59, 640.

7. Hawthorne, S. B.; Miller, D. J. Chromatogr. 1987, 403, 63.

8. Pellizzari, E. D.; Damien, B.; Schindler, A.; Lam, K.;
    Jeans, W. Preparation and Evaluation of New Sorbents for
    Environmental Monitoring, Volume 1, Final Report on EPA project
    68-02-3440, 1982.

9. Parcher, J. F.; Selim, M. I. Anal. Chem. 1979, 51, 2154.

10. Parcher, J. F.; Johnson, D. M. J. Chromatogr. Sci. 1985, 23,
    459.

RECEIVED September 29, 1988

SPECIAL APPLICATIONS

Chapter 21

Inverse Gas Chromatography of Coals and Oxidized Coals

P. H. Neill and R. E. Winans

Chemistry Division, Argonne National Laboratory, 9700 South Cass Avenue, Argonne, IL 60439

Inverse gas chromatography using methane as a probe is used to determine the physical and chemical alterations that coals and oxidized coals undergo as they are heated from 50 to 450°C. At temperatures less than 120°C retention is dominated by water in the pore structure of the sample. In the intermediate temperature region between 120°C and 350°C pyrolysis high resolution mass spectrometry is used to assign the small transitions that are observed to the loss of small amounts of volatile matter. Giesler plastometry and microdilatometry results are used to show that massive disruption of the coal structure occurs at temperatures above 350°C. The starting temperature and the sorption enthalpy are found to vary in a linear manner with the carbon content and degree of oxidation of the coals in this region.

Previous papers (1-4) have documented results relating to the application of inverse gas chromatography (IGC) to coals and air oxidized coals. This paper summarizes results pertaining to the reproducibility of this IGC application, documents the use of IGC in monitoring the effect of air oxidation on coal fluidity, and compares the results via those obtained with pyrolysis mass spectrometry and Giesler plastometry.
    The objective of this study is to develop IGC as a method of elucidating the chemical and physical changes that coals undergo in varying storage environments. It is known that minor changes in the structure of coal can affect changes in plasticity when a coal is heated. However, the techniques traditionally used to measure plasticity, such as Giesler plastometry and dilatometry, do not yield information that can be used to understand the underlying chemical and physical changes in the coal structure. It is believed that IGC shows promise in providing missing information.

0097–6156/89/0391–0290$06.00/0
© 1989 American Chemical Society

Two general categories of change are of special interest. The first category is the effect of weathering on high temperature fluidity. It is known that minor oxidation of the coal drastically decreases it's ability to undergo a fluid transition (5,6). The reason for this is not known. One widely held belief is that the oxidation process significantly increases the cross-link density and simultaneously removes labile hydrogen, which is needed to stabilize the free radicals formed in the fluidization process. The second category includes the possible non-oxidative changes that may occur when a coal is stored for a long time in an inert atmosphere. Since pristine samples were unavailable in the past, there is no direct evidence that these changes occur. However, one could speculate that since coal is a viscoelastic material that was formed and stored for millions of years under great physical stress, the release of this stress would allow a slow migration to a lower energy configuration. This change in physical structure may be observable in the IGC results.

In contrast to classical GC, IGC probes the stationary phase, in this case a coal, by determining the retention volume of known compounds. The term inverse chromatography was first applied by Davis and Peterson (7), who utilized IGC to determine of the degree of oxidation of asphalt. In later work, the experiments were extended to include measurements of thermodynamic properties (8). Because of their results with oxidized asphalt, IGC may be a good candidate for investigating coals.

IGC, although a relatively new method, has been widely applied in the characterization of polymers (9-13). Specific applications include measurement of glass transition temperatures (14), the degree of crystallinity (15), melting point (16), thermodynamics of solution (17) and chemical composition (7,8). IGC has also been applied to dried coals at temperatures less than 85°C to determine the enthalpy for sorption of methane and oxygen (18). Methane sorption enthalpies ranged from 4.4 to 0.9 Kcal/mole for Czechoslovakian coals containing 90.7 to 83.3% carbon on a dry, ash-free basis.

Two types of information are provided by the IGC experiment. The slope of a plot of the log of the retention time versus the inverse of the temperature is proportional to the enthalpy of retention for the probe molecule on the coal. This is a thermo-dynamic measure of the strength of the interaction between the probe molecule and the coal. The temperatures at which major changes in slope are observed represent the points where the mechanism of retention has changed, indicating that a significant change in the chemical or physical structure of the coal has occurred.

Materials and Methods

The compositions of each coal used in this study is shown in Table I. The conditions under which these pristine samples were collected and prepared have been reported previously (19). The weathered Upper Freeport sample was exposed to air and sunlight at room temperature for four weeks, while the oxidized coal was heated

at 100°C for 90 h. All of the -100 mesh samples were thoroughly
mixed with non-porous -60 to -400 mesh glass beads to give a mixture approximately of 10 wt-% coal. A blank experiment utilizing
only glass beads yielded an effective slope of zero, indicating
that adsorption of methane on the non-porous glass beads was minimal. The six foot by 1/4 inch glass columns were packed with
approximately 30 grams of the mixture. All transfers and weighings
were performed in a glove box under a nitrogen atmosphere.

Table I. Composition of Coals Studied

Coal	% C dmmf*)	Empirical formula
Upper Freeport mv Bituminous (APCS #1)	85.5	$C_{100}H_{66}N_{1.5}S_{0.3}O_{6.6}$
Illinois Herrin seam hvC Bit. (APCS #3)	77.7	$C_{100}H_{77}N_{1.5}S_{1.2}O_{13.0}$
Pocahontas lv Bituminous (APCS #5)	91.0	$C_{100}H_{59}N_{1.2}S_{0.2}O_{2.0}$
Bruceton hvA Bituminous	82.3	$C_{100}H_{78}N_{1.6}S_{0.3}O_{9.3}$
Oxidized Upper Freeport	72.9	$C_{100}H_{73}N_{1.5}S_{1.2}O_{20.0}$

* Dry, Mineral-Matter-Free Basis

    A diagram of the experimental apparatus is provided in Figure
1. The Perkin-Elmer GC was equipped with a single flame ionization
detector and was capable of operating at temperatures below 450°C.
The injector was a computer controlled Carle gas sampling valve in
a thermostated box. Flow control was provided by two flow con-
trollers the first with a 0 to 5 mL/min element and the second with
a 0 to 60 mL/min element. The second controller was connected to
the injector through a computer-controlled solenoid. All facets of
the experiment were controlled and the data analyzed, using an IBM
PC computer.
    A flow diagram of the IGC experiment is shown in Figure 2.
The experiment began with an initial equilibration period of at
least 24 h, during which the flow rate of the helium carrier gas
was cycled between it's chromatographic rate of 1 mL/min and the
higher rate (30 mL/min). The higher flow rate was used to speed
the removal of volatile matter from the column after each temperature increment. The probe, 10% methane in argon, was injected

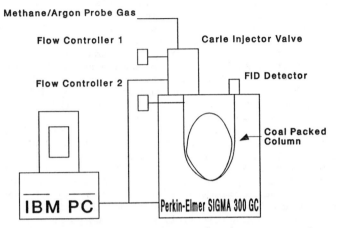

Figure 1.   Experimental apparatus for inverse gas chromatography experiment.

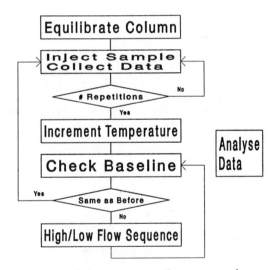

Figure 2.   Flow diagram of inverse gas chromatography experiment.

into the column and its retention time determined at a specific temperature. In the experiment, duplicate chromatograms were obtained in four degree increments between 50 and 450°C. After each temperature increment, the baseline signal was determined and compared to the baseline at the previous temperature. If it had increased more than 10%, the solenoid isolating the second flow controller was opened, thereby increasing the rate at which the coal volatile matter was purged from the column. Even with increased flow rate, an IGC experiment could take up to a month. The injector was held at 100°C and the detector at 400°C.

The pyrolysis high resolution mass spectrometry (PyHRMS) technique has been described in detail previously (20). Briefly, the coal sample was placed on a platinum rhodium mesh on the end of a probe as a slurry. After the solvent had evaporated, the probe was inserted into the mass spectrometer and positioned within 5 mm of the source. The probe, which had been previously calibrated with an infra-red thermometer, was computer-controlled to give a temperature profile beginning at 100°C and increasing at 50°C/min to 800°C. The precise masses were matched to their corresponding chemical structures by computer programs developed in-house. This technique results in the relatively slow vacuum pyrolysis of the coal sample.

Giesler plasticity measurements on the coals were obtained using the ASTM procedure D2639-74.

Results and Discussion

The results from the IGC experiment on Upper Freeport Coal (APCS No. 1) are presented in Figure 3. There are three retention mechanisms likely to affect the IGC results. These mechanisms include molecular sieving, surface adsorption, and the dissolution of the probe in the stationary phase. The observed retention behavior results from the combination of the effects operating under the given conditions. The specific retention volume, V_g, is given by the sum of the retention terms in Equation 1:

$$V_g = K_d V_L + K_o A_L + K_s WL, \tag{1}$$

where K_d, K_o, and K_s are the partition coefficients for gas-pore, gas-solid, and gas-liquid partition, respectively; V_L is the accessible pore volume; A_L is the accessible surface area; and W_L is the accessible mass of the stationary phase. It is primarily the physical state of the stationary phase that determines the retention mechanism. The retention volume is related to the free energy or enthalpy of adsorption as follows:

$$\Delta H_a \propto \delta \ln V_g / \delta (1/T), \tag{2}$$

Thus, plot of log V_{ret} versus $1/T$ gives a slope proportional to H_a. Changes in the thermodynamics of retention result in the discontinuities in the curve in Figure 3.

The plot of the log of the retention time versus the inverse of the temperature in Kelvin is divided into three general regions based on the slope of the curve. The first major transition indicated by the significant change in slope is normally found between 100 and 120°C. It is believed that this transition is caused by the dehydration of the pore structure of the coal. This change in slope indicates that the way the methane probe molecule interacts with the wet coal, or the mechanism of retention, is different enough from that of the dry coal to produce the observed deviation in the curve. In the intermediate temperature region (approximately 115° to 350°C) a relatively constant slope is observed, indicating that the retention mechanism for the methane probe molecule does not change significantly. However, above 200°C, a large number of small transitions are encountered that do not result in a major deviation from the prevailing slope. It is believed that these deviations are due to the loss of volatile matter occluded within the coal structure. The loss of relatively minor amounts of material should result in small deviations in retention time, which are observed but would not be expected to modify the overall mechanism of retention significantly. At temperatures greater than 350°C, there is a marked change in the slope, indicating a major change in the mechanism by which the methane probe is retained by the coal.

Other researchers using a variety of methods have observed transitions. For example, differential scanning calorimetry studies by Mahajan et al. (21) showed two endotherms in the 300 to 400°C range that appear at increasing temperatures with increasing coal rank. Patrick, Reynolds, and Shaw (22) observed a transition to completely anisotropic material in the temperature range between 370 and 420°C in optical anisotropy experiments using vitrains with carbon contents <89%. In addition, several researchers (23) have reported that, after heat treatment, both the amount of pyridine-soluble material and the average molecular weight of the material reach a maximum between 300 and 350°C. This also corresponds to the temperature range in which Giesler Plastometer results indicate that the coal is in a fluid state.

The reproducibility of the IGC technique is shown in Figure 4, where the results from duplicate experiments using Illinois Herrin Seam hvcB are presented. The results from the two experiments are similar. Not only do the major transitions occur at the same temperatures, but many of the minor transitions are reproduced as well. The two curves appear to be most dissimilar in the low temperature region. The low temperature region is where it is believed that changes in retention are dominated by water in the coal; thus, the results are most sensitive to variation in sample handling. The minor transitions in the intermediate temperature region (120 to 300°C) are reproduced accurately. Both curves show minor transitions at approximately 190, 230, and 285°C. The temperature of the major transition appears to be similar for the two experiments. An exact assignment of the temperature of this transition in experiment two is impossible because of a problem with the data system, which resulted in the loss of data between 300 and 326°C. In the higher temperature region (>326°C), reproducible transitions occurred near 380 and 420°C. The enthalpies of sorption were very similar and within the calculated precision of the data, as shown in Table II.

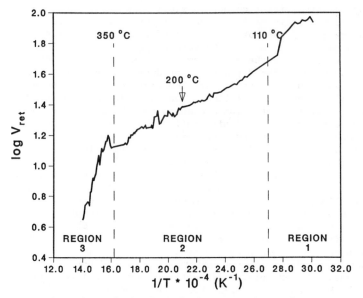

Figure 3. Plot of the log of the retention volume versus the inverse of temperature for Upper Freeport coal.

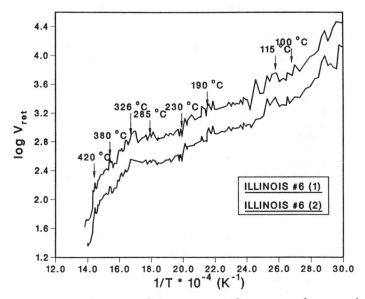

Figure 4. Comparison of inverse gas chromatography experiments on Illinois No. 6 coal.

TABLE II. Enthalpies of Sorption in Kcal/mole for the Three
Temperature Regions[+]

Coal	50 to 100°C Very Low Temp. Region	100 to 300°C Low Temp. Region	350 to 450°C High Temp. Region
Upper Freeport mv	2.9±0.5	1.93±.07	2.5±0.4
Upper Freeport (weathered)	4.0±0.2	1.97±.04	7.7±0.4
Upper Freeport (oxidized)	3.5±0.2	1.98±.02	4.9±0.2
Pocahontas lv	----- *	2.01±.04	14.8±0.4
Bruceton hvA	1.6±0.2	1.96±.04	7.7±0.4
Illinois Herrin (1)	4.0±0.3	2.00±.16	4.9±0.5
Illinois Herrin (2)	4.3±0.3	1.90±.15	5.1±0.5

+   Temperatures indicated are approximate; the actual boundaries
    were determined by the temperature of the actual transitions in
    the specific experiment.

*   Pocahontas coal exhibited a negative slope in this region for
    some unknown reason.

The temperature at which the major transition peak occurs for
each of the coals is plotted against the percent carbon in Figure
5. This peak marks the start of the high temperature region in
each of the experiments. This transition can be seen in Figure 3
just above 350°C for the Upper Freeport coal; in Figure 4, just
below 326°C for Illinois Herrin seam hvcB; and in Figure 7 at
approximately 370°C for the Pocahontas coal. The limited number of
samples studied precludes stating that a correlation exists; however one is safe in assuming that a trend is indicated.

The enthalpy of sorption for each of the coals plotted versus
carbon content is presented in Figure 6. The data were derived
from a least squares linear regression of the IGC data taken above
the temperature of the major transition. As with the temperature
of the major transition, a trend is definitely indicated. However,
several of the enthalpies are unrealistically high. This is probably due to loss of volatile matter from the coal in this temperature region. Since the actual mass of coal in the column at each

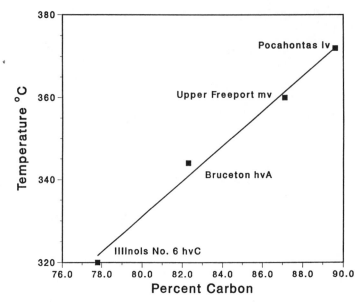

Figure 5. The temperature of the start of the major transition versus carbon content (dry mineral matter free basis).

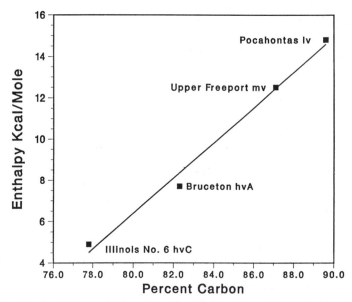

Figure 6. The enthalpy for the high temperature region (350 to 450°C) versus carbon content (dry mineral matter free basis).

injection temperature could not be determined, V_g could not be calculated, leading to the erroneously high value of ΔH.

Figure 7 compares the results from the IGC experiment with results obtained from PyHRMS for Pocahontas coal. Although the pyrolysis was started at 100°C, the total ion current signal (TIC) below 208°C was dominated by an internal standard added to the coal. Other experiments indicate that little material of molecular mass >31 amu is released below this temperature. Above 208°C minor transitions in the IGC curve become numerous. This indicates that these minor transitions in the IGC are represent losses of volatile material. The maximum TIC is observed at a temperature close to that of the major transition, in the IGC experiment. At temperatures above the major transition the TIC curve indicates that the release of volatile material begins to decrease significantly.

A comparison between the IGC results for Upper Freeport coal with results obtained via Giesler plastometry is presented in Figure 8. The marked increase in retention volume that indicates the start of the third region correlates well with the initial softening temperature. For the coals investigated thus far, the transition is observed in the IGC 10 to 15°C before the initial softening temperature.

Microdilatometer results can also be correlated with a non-pristine coal, Bruceton hvA bituminous (24). As shown in Figure 9. the softening temperature (T_s) corresponds to a second minor transition within the high temperature region. Contraction temperature (T_c) occurs at the steepest part of the high temperature transition curve in the IGC.

Figure 10 illustrates the effect of air oxidation on the IGC pattern for Upper Freeport coal. Weathering and high temperature oxidation do not seem to have much effect on the lowest temperature transition, although it may be reduced in intensity by the action of the two processes. This is understandable if one considers the fact that both the oxidized and weathered samples have been exposed to significantly less humidity for varying periods of time. The intermediate temperature region is similar for each of the four samples, as evidenced by the heat of adsorption data presented in Table II. For all practical purposes, the heat data reported are the same for all the samples in this temperature range. The main differences in the IGC for these three samples are in the high temperature region where the most drastic chemical changes are expected to occur. This region is expanded in Figure 11. The unoxidized sample shows a pattern similar to that observed with other non-oxidized samples, which indicates that the coal has become fluid. The weathered sample seems to undergo the same transition. However, the heat of interaction between the probe molecule and the coal is reduced significantly. Thus, one can conclude that the degree to which the transition occurs has been reduced by weathering. This conclusion agrees with the generally held belief that even minor oxidation of the coal increases the degree of cross-linking in the coal structure, reducing the coal's ability to become fluid.

The most drastic change in this region is observed in the oxidized sample. There is little change in the heat of interaction

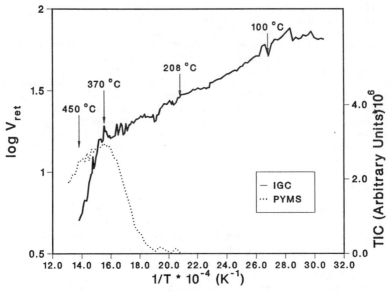

Figure 7. Comparison of inverse gas chromatography and pyrolysis high resolution mass spectrometry experiments for Pochontas
lv bituminous coal.

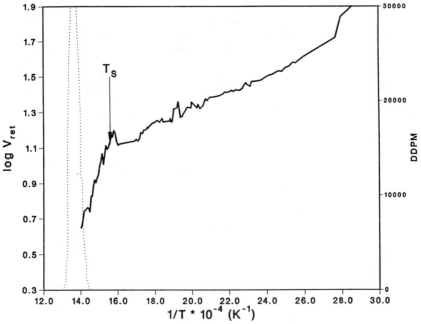

Figure 8. Inverse gas chromatography of Upper Freeport Mv coal
compared with the Geisler plastometer results from the same
coal (_____ IGC, ... Geisler).

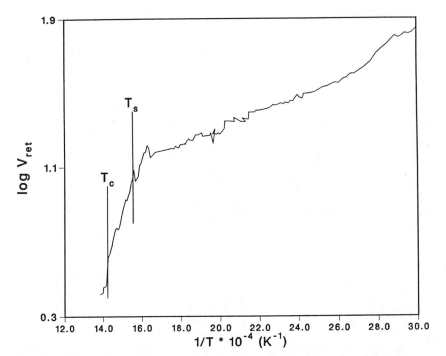

Figure 9. Comparison of inverse gas chromatography data from the Bruceton coal with microdilatometer measurements of a coal from the same seam (data from ref. 24).

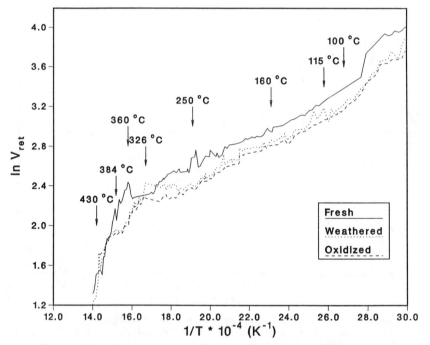

Figure 10. Comparison of inverse gas chromatography experiments on the Upper Freeport coals that have undergone various oxidative treatments.

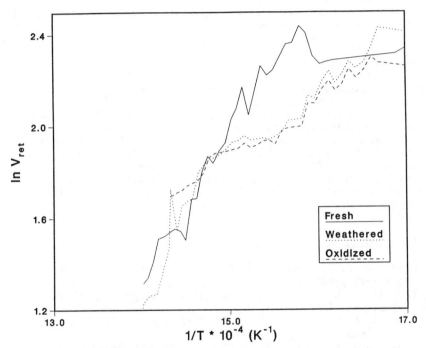

Figure 11.   Expanded view of the high temperature region in
Figure 10.

between the intermediate temperature region and the high temperature region (3 kcal/mole). This result indicates that the cross-
link density was raised high enough that the coal no longer undergoes a fluid transition.
     Overall, IGC appears to be a reproducible method for following
the chemical and physical changes that occur when coals are heated
in an inert atmosphere. Differences in the transition temperature
and enthalpies of sorption can be observed for coals of various
rank. PyHRMS results indicate that the minor transitions observed
in the intermediate temperature region are a result of the loss of
volatile matter from the coal.

Acknowledgments

The coal samples (excluding Bruceton) were provided by the Argonne
Premium Coal Sample Program. The coal Geisler plasticity measure-
ments were performed by P. C. Lindahl of the Argonne Analytical
Chemistry Laboratory. The Bruceton coal was provided by
J. W. Larsen of the Department of Chemistry, Lehigh University.
Reference 20 was translated from Czechoslovakian by Petr Vanýsek of
the Department of Chemistry, Northern Illinois University. This
work was performed under the auspices of the Office of Basic Energy
Sciences, Division of Chemical Sciences, U. S. Department of
Energy, under contract number W-31-109-ENG-38.

Literature Cited

1. Winans, R. E.; Goodman, J. P.; Neill, P. H.; McBeth, R. L.
   ACS Fuel Chemistry Division Preprints, 1985, 30(4), 427.
2. Neill, P. H.; Winans, R. E. ACS Fuel Chemistry Division
   Preprints, 1986, 31(1), 25.
3. Neill, P. H.; Winans, R. E. ACS Fuel Chemistry Division
   Preprints, 1987, 32(4), 266.
4. Winans R. E.; Neill, P. H. Proc. of the Coal Research
   Conference, Wellington, New Zealand, 1987, p. R-8.3.
5. Senftle, J. T.; Davis, A. Int. J. Coal Geol., 1984, 3, 375.
6. Huffman, G. P.; Huggins, F. E.; Dunmyre, G. R.; Pignocco, A.
   J.; Lin, M. C. Fuel, 1985, 64, 849.
7. Davis, T. C.; Peterson, J. C. Analytical Chemistry, 1966,
   38, 1938.
8. Dorrence, S. M.; Peterson, J. C. Analytical Chemistry, 1969,
   41, 1240.
9. Gilbert, S. G. in Advances in Chromatography,   Giddings, J.
   C.; Grushka, E.; Cazes, J.; Brown, P. R. Eds.; Marcel
   Dekker: New York 1984; Vol. 23, p. 199.
10. Berezkin, V. G.; Alishoyev, V. R.; Nemirovskaya, I. B. J.
    Chromatography Library 1973, 10, 197.
11. Guillet, J. E. in New Developments in Gas Chromatography;
    Purnell, J. H. Ed.; Wiley: New York, 1973; p. 178.
12. Card, T. W.; Al-Saigh, Z. Y.; Munk, P. Macromolecules, 1985,
    18, 1030.

13. Munk, P.; Al-Saigh, Z. Y.; Card, T. W. Macromolecules, 1985, 18, 2196.
14. Smidsrod, O.; Guillet, J. E. Macromolecules, 1969, 2, 272.
15. Braun, J. M.; Guillet, J. E. Macromolecules, 1976, 9, 349.
16. Alishoyev. V. R.; Berezkin, V. G.; Mel'nikova, Y. V. J. Phys. Chem., 1965, 39, 105.
17. Varsano J. L.; Gilbert, S. G. J. Pharm. Sci., 1973, 62, 187, 1192.
18. Taraba, B.; Čáp, K. Uhli, 1985, 33(7-8), 278.
19. Vorres K.; Janikowski, S. ACS Fuel Chemistry Division Preprints, 1987, 32(1), 492.
20. Winans, R. E.; Scott, R. G.; Neill, P. H.; Dyrkacz, G. R.; Hayatsu, R. Fuel Processing Technology, 1986, 12, 77.
21. Mahajan, O. P.; Tomita, A.; Walker, P. L., Jr., Fuel, 1976, 55, 63.
22. Patrick, J. W.; Reynolds, M. J.; Shaw, F. M. Fuel, 1973, 52, 198.
23. Marsh, H.; Stadler, H. P. Fuel, 1967 46, 351, and references therein.
24. Khan, M. R.; Jenkins, R. G. Fuel, 1984, 63, 109.

RECEIVED December 27, 1988

Chapter 22

Modified Frontal Chromatographic Method for Water Sorption Isotherms of Biological Macromolecules

Seymour G. Gilbert

Food Science Department, Cook College, Rutgers, The State University, New Brunswick, NJ 08903

The conventional inverse gas chromatography (IGC) is based on equations that assume equilibrium is established during the course of the chromatograph. Consequently, those stationary phases that exhibit marked hysteresis in sorption/desorption give IGC sorption data at considerable variance with long-term gravimetric methods. A modified frontal procedure was developed that avoids the assumption of equilibrium to enable studies of interaction kinetics of gas phase components with a stationary phase, such as a biopolymer, having entropic as well as enthalpic relations affected by concentration shifts and time dependent parameters.

Methods for determining sorption isotherms by gas chromatography have been published by various authors([1-5]). The methods used have been elution and frontal chromatography. The first combines sorption and desorption so that any hysteresis in the equilibrium transport from gas to stationary phase and back to the gas phase can produce corresponding errors. The Kiselev-Yashin equation, as shown in Figure 1,

$$A = (Mp/m)(x/y) \qquad (1)$$

A = partition in g/g solute/solvent
Mp = mass solute input
m = mass solvent
x = prepeak area
y = peak area

does not necessarily correct for non-linear, non-equilibrium chromatography, despite the partial cancellation of errors in desorption by use of prepeak area ratio to peak area ratio. Since time is part of the area calculation, any diffusional or other kinetic factors may induce errors ([6]).

The use of frontal chromatography would appear to avoid the errors produced by hysteresis in that the sorption and desorption processes can be separated in time. An extensive study of this method for water sorption, using freeze-dried coffee as a

0097–6156/89/0391–0306$06.00/0

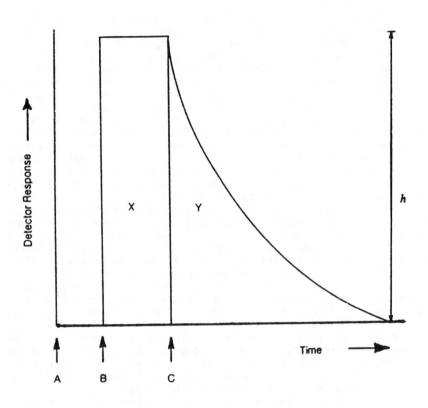

where:

A = point of injection
B = point of emergence of unsorbed peak (air)
C = point of emergence of probe peak
X = prepeak area
Y = peak area
h = height of peak

Figure 1. Illustration of Pulse Analysis.

stationary substrate, was conducted by Apostolopolous and Gilbert
(7), and a study on desorption on starch and sugars was conducted
by Paik and Gilbert (8).

Paik and Gilbert found discrepancies between frontal IGC for
sorption compared to a static method that used a long-term water
sorption from saturated salt solutions (Figure 2). They also showed
that if a sufficiently high temperature rise was used, following
apparent complete desorption, the area formed from the temperature
pulse adds the strongly bound water to the eluate to provide a
correction factor for the difference between water sorption by the
static weighing and dynamic IGC frontal methods (Figure 3).
Apostolopolous and Gilbert (7) showed that the divergence was greatest at the lowest temperatures and lowest mass absorbed (maximum
bound water conditions), and least divergent at high temperature
and high mass sorbed. Gilbert (9) designated the high water content
region at low temperature as the clustered water region identifiable with free water, but not with capillary water unless solutions
of low molecular weight solutes are present.

In this case, the cluster integral of Zimm and Lundberg (10)
for water-water interactions is high enough to approach the vapor
pressure of pure water as a limit and fugacity equal to 1.

At lower water contents, the water vapor pressure is the sum
of the fugacities of water at all sites. This sum includes differing enthalpies of the first water molecule sorbed and that of any
clusters of water at such sites. The same fugacity average can be
obtained from a number of combinations of the degree of heterogeneity, the frequency distribution of such enthalpies and total
number of such sites per unit of solid phase.

Desorption will differ from sorption in proportion to the
degree and distribution of such heterogeneity of enthalpies since
the entropic relations are different in a solid, depending on the
direction of the concentration gradient as it affects the kinetic
factors of density and diffusivity within the matrix.

These considerations are particularly important for non-linear
or concentration dependent relations and non-equilibrium conditions, such as those found in chromatographic systems showing
markedly skewed peaks (6). As these authors have shown, there is no
identifiable solution to the problem of the thermodynamic properties of the highly skewed chromatogram peak. Thus, the elution
method is only valid for equilibrium chromatography.

Biologically derived macromolecules are highly heterogeneous
in site distribution from composition differences in monomers and
where amorphous/crystalline regions are present. The differences
previously found in sorption are related to the difference in
equilibrium time for static and chromatographic methods (8).

A method to circumvent this dilemma was sought by Ferng (11).
An extensive and detailed study of static sorption methodology was
first conducted to provide a basis of reproducible data for starches of different macromolecular structure. This was followed by studies of sorption isotherms by IGC with different GC conditions
including zero loads with empty and supposedly inert support
material (diatomaceous earth). The data showed that the response of
the thermal conductivity detector (TCD) to controlled chromatographic conditions of temperature, flow rate, and partial water vapor

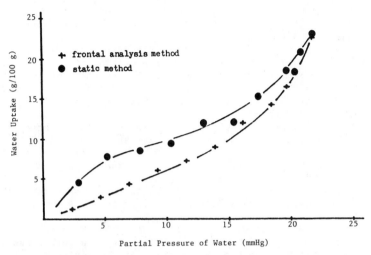

Figure 2. Sorption Isotherm of Corn Starch Determined at 25°C by Static Method.

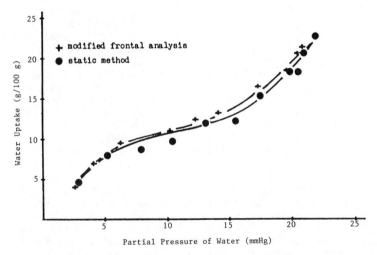

Figure 3. Sorption Isotherm of Starch Determined at 25°C by Modified Frontal Analysis and Frontal Method.

concentration was highly reproducible and linear with mass and
temperature. Thus, the area response discrepancies were produced by
incomplete elution. This conclusion was reinforced by using a pulse
of high temperature to desorb the water from sites of high negative
enthalpy of binding (Figure 4). Under these conditions, the mass/
area ratio (Ka) of eluted solute of empty, blank and substrate
loaded columns were equal (Figure 5).

When coated or highly dispersed starch substrates were used to
reduce diffusivity effects, the mass/area ratio of the prepeak
region of the chromatogram was equal to that of the peak area in
the low mass region, reaching a plateau with no change in area as
the substrate became saturated for a highly dispersed, freeze
dried high amylose starch.

When three dimensional particles were used instead of coatings, the diffusional and structural kinetics showed marked differences in the rates of mass to area in the concentration dependent
regions of partition coefficient (lower temperature and mass
sorbed). Thus, the fundamental assumption of the Kiselev Yashin
equation that the mass/area ratio was equal for prepeak and peak
regions was not met, and agreement with static data was fortuitous
at best.

The same considerations apply to the prepeak area of the
frontal method. The eluted peak represents the unabsorbed mass,
which is proportional to the integration of the advancing front
over elution time. However, there is an error from incomplete elution reducing the prepeak area, since the response height was
asssumed to be equal to the water vapor pressure in equilibrium
with a specific amount of water in the solid phase in the derivation of equation (1). The error then is proportional to the hysteresis, since the water in the solid phase is equivalent to a higher
vapor pressure if equilibrium has been attained. A mass balance
approach was developed that used a defined input mass with the non
sorbed water mass calculated from the front peak area and sorption
by difference (12). This produced agreement at the lower concentration stage of the isotherm since the uneluted mass was accounted
for. The difficulty was that a large number of mass increments was
required to determine a strong non-linear (for example, sigmoid)
isotherm. This increased the operating time and complexity of the
method, as did the Paik and Gilbert procedure (8).

Since the product of flow rate, time and concentration equal
the input mass, a constant input concentration permits the calculation of mass from either time or retention volume. Empty columns
provide an essentially constant ratio of input mass to time at
constant flow rate with the concentration of water vapor fixed by
the temperature of the carrier gas saturated with water vapor (100%
RH or Aw of 1 see Figure 6). This state can be achieved with
substrates that do not dissolve in water when saturated (for
example, starches and many proteins), or when the relative humidity
is constant but insufficient to allow uptake to produce a highly
multilayered or clustered water state in the substrate equivalent
to a continuous water phase or solution. This condition requires a
source of humidified gas as in (7). The sorption isotherm equation
is then given by

$$A = (t_c Kt - Y_c Ka)/m \qquad (2)$$

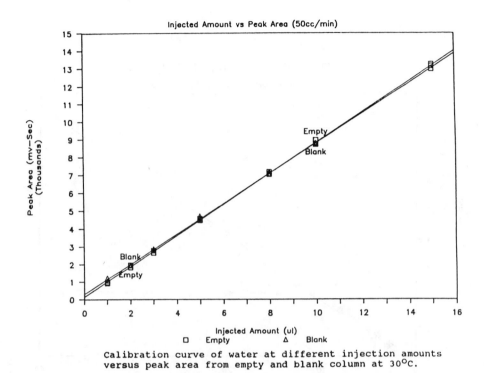

Calibration curve of water at different injection amounts versus peak area from empty and blank column at 30°C.

Figure 4. Concentration Profile Formed During Operation of Modified Frontal IGC.

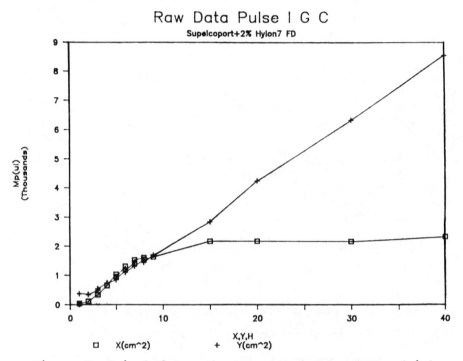

Figure 5. Injected Amount versus Peak Area (50 cc/min).

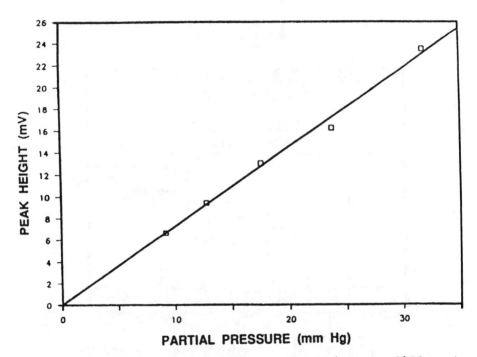

Figure 6. Partial Pressure versus Peak Height (5 different temperatures, 10, 15, 20, 25, 30 and 35°C).

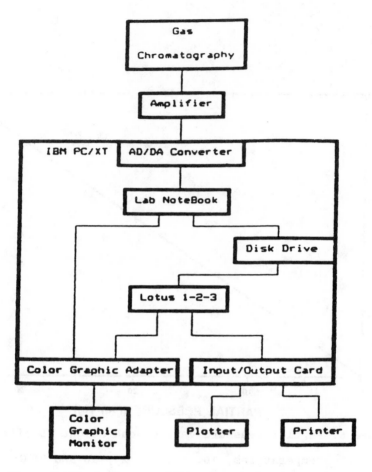

Figure 7. Block Diagram of Data Acquisition System.

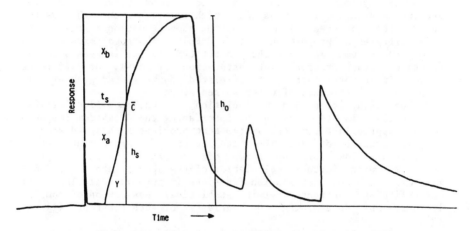

Figure 8. Modified Frontal by Pulse Method.

Pc = (hc/ho)Po or (hc/ho) = Aw
where: Ka = (mass/area) constant
          Kt = (mass/time) constant
          tc = time at defined height hc = f(Pl)
          Yc = area of peak at tc
          Pc = water vapor pressure at tc
          ho = maximum height at saturation = f(Po)
          Po = saturation vapor pressure of water
          Pl = equilibrium   vapor pressure at a specific water
               content in the solid phase
          hc = response height at a specific frontal time, tc, and
               peak area, Yc.
          A = mass of sorbed solute (water) in stationary phase at
              time, tc.

When hysteresis produces a water vapor pressure (Pd) for
desorption at a specific mass (A), which is greater than the observed Pc, the peak area calculated by the above equation represents an
area difference proportional to (Pd-Pc)tc. The total mass entered
into the column at tc is the product tcKt, which increases to the
saturated state response at saturation or Aw=1. The difference
between this total mass and that related to the eluted peak area is
given above as the mass of water sorbed at Pc.

The acquisition and manipulation of the modified frontal
chromatogram is facilitated by interfacing the GC system response
to an appropriate computer system that provides high speed storage
and integration of data from the detector. An appropriate set of
algorithms allow for conversion of the raw data into sorption
isotherms with further calculation of the appropriate thermodynamics and cluster distribution functions (Figure 7). Calculation of
curve-fitting constants permit statistical evaluation of complex
isotherms by determining the variance of such constants (11 and
12).

A complete sorption isotherm is obtained within two hours with
this combination of GC and microprocessor. If the substrate is
stable under the GC conditions, a family of isotherms can be produced within two days or within one if automated to operate on a
24-hour day. A represen-tative chromatogram is shown in Figure 8.

Kinetic factors can be present and dominate sorption rate
(desorption of crystal structure, chain folding or unfolding, and
structural shifts on swelling). In such cases the discrepancies
between static and IGC methods reflect the time-related difference
and not concentration dependency of the isotherm (13). The suggested procedure will not resolve such differences unless sufficiently
low flow rates are used. These discrepancies are of great importance in time-related studies, including storage life stability
under different environments, as provided by packaging systems of
different permeability to water vapor. Some of the possible applications of this new method were discussed in another study (14).

Acknowledgments

Paper   number D-10535-8-88 of the Journal Series, New Jersey
Agricultural Experiment Station, Cook College, Rutgers The State
University, Department of Food Science, New Brunswick, New Jersey

08903. This work was supported in part by State funds and the Center for Advanced Food Technology (CAFT), a New Jersey Commission on Science and Technology Center.

Literature Cited

1. Gregg, S.J.; Stock R. (1958). In Gas Chromatography D.H. Desty, Ed. pg. 90, Butterworth, London.
2. Conder, J.R.; Purnell J.H. Trans. Faraday Soc. 1969, 65. 824-848.
3. Eberly, P. E. Jr. J. Phys. Chem. 1961, 65, 1261-1265.
4. Kiselev, A.Y.; Yashin Ya.I. (1969). "Gas Adsorption Chromatography." Plenum Press, New York.
5. Gray, D.G.; Guillet, J.E. Macromolecules 1972 ,5, 1972, 316-320.
6. Conder J.R.; McHale, S.; Jones, M.A. Analyt. Chem. 1986, 58, 2663.
7. Apostolopoulos, D.; Gilbert, S.G. (1984). Instrumental Analysis of Foods. Vol 2. G. Charalombous and G. Inglett, eds. Academic Press. New York.
8. Paik, S.W.; Gilbert, S.G. J. Chromatography. 1986, 351-417.
9. Gilbert, S.G. in The Shelf Life of Foods and Beverages, ed. G. Charalambous. Elsevier Science Publishers B. V.,Amsterdam.
10. Zimm, B.H.; Lundberg, J.L. J. Phys. Chem. 1956, 60, 425-428.
11. Ferng A.L. (1987). A study of Water Sorption of Corn Starch from Various Genotypes by Gravimetric and Inverse Gas Chromatographic Method, Ph.D. Thesis. Rutgers University. New Brunswick, New Jersey.
12. Gilbert, S.G.; Ferng, A.L. 1987. New method for Sorption Isotherms by Gas Chromatography. 47th Annual IFT Meeting Las Vegas, Nevada.
13. Il, Barbara; Daun, H.; Gilbert, S.G. 1987. Water Sorption of Gliadin. J. Food Science, paper in preparation.
14. Gilbert, S.G. 1988. Applications for Research in Kinetic and Thermodynamic Problems in Food Science. ACS Symposium, Toronto, Canada.

RECEIVED December 5, 1988

INDEXES

Author Index

Affiliation Index

Subject Index

Production by Paula M. Bérard and Rebecca A. Hunsicker
Indexing by A. Maureen Rouhi

Elements typeset by Hot Type Ltd., Washington, DC
Printed and bound by Maple Press, York, PA

Other ACS Books

Biotechnology and Materials Science: Chemistry for the Future
Edited by Mary L. Good
160 pp; clothbound, ISBN 0–8412–1472–7, paperback, ISBN 0–8412–1473–5

Chemical Demonstrations: A Sourcebook for Teachers
Volume 1, Second Edition by Lee R. Summerlin and James L. Ealy, Jr.
192 pp; spiral bound; ISBN 0–8412–1481–6
Volume 2, Second Edition by Lee R. Summerlin, Christie L. Borgford, and Julie B. Ealy
229 pp; spiral bound; ISBN 0–8412–1535–9

The Language of Biotechnology: A Dictionary of Terms
By John M. Walker and Michael Cox
ACS Professional Reference Book; 256 pp;
clothbound, ISBN 0–8412–1489–1; paperback, ISBN 0–8412–1490–5

Cancer: The Outlaw Cell, Second Edition
Edited by Richard E. LaFond
274 pp; clothbound, ISBN 0–8412–1419–0; paperback, ISBN 0–8412–1420–4

Chemical Structure Software for Personal Computers
Edited by Daniel E. Meyer, Wendy A. Warr, and Richard A. Love
ACS Professional Reference Book; 107 pp;
clothbound, ISBN 0–8412–1538–3; paperback, ISBN 0–8412–1539–1

Practical Statistics for the Physical Sciences
By Larry L. Havlicek
ACS Professional Reference Book; 198 pp; clothbound; ISBN 0–8412–1453–0

The Basics of Technical Communicating
By B. Edward Cain
ACS Professional Reference Book; 198 pp;
clothbound, ISBN 0–8412–1451–4; paperback, ISBN 0–8412–1452–2

The ACS Style Guide: A Manual for Authors and Editors
Edited by Janet S. Dodd
264 pp; clothbound, ISBN 0–8412–0917–0; paperback, ISBN 0–8412–0943–X

Personal Computers for Scientists: A Byte at a Time
By Glenn I. Ouchi
276 pp; clothbound, ISBN 0–8412–1000–4; paperback, ISBN 0–8412–1001–2

Chemistry and Crime: From Sherlock Holmes to Today's Courtroom[A
Edited by Samuel M. Gerber
135 pp; clothbound, ISBN 0–8412–0784–4; paperback, ISBN 0–8412–0785–2

For further information and a free catalog of ACS books, contact:
American Chemical Society
Distribution Office, Department 225
1155 16th Street, NW, Washington, DC 20036
Telephone 800–227–5558